CENTRIFUGAL AND AXIAL COMPRESSOR CONTROL

CENTRIFUGAL AND AXIAL COMPRESSOR CONTROL

Gregory K. McMillan

MOMENTUM PRESS

Momentum Press, LLC., New York

First published in 2010 by
Momentum Press®, LLC
222 East 46th Street, New York, N.Y. 10017
www.momentumpress.net

ISBN-13: 978-1-60650-173-3 (softcover)
ISBN-10: 1-60650-173-9 (softcover)

ISBN-13: 978-1-60650-174-0 (e-book)
ISBN-10: 1-60650-174-7 (e-book)

DOI forthcoming

Cover Design by Jonathan Pennell
Interior Design by Scribe, Inc. (www.scribenet.com)

First Edition March 2010

10 9 8 7 6 5 4 3 2 1

Printed in Taiwan R.O.C.

Contents

Figures

Tables

Introduction to the Student

1-1 CONTENTS

The cost of machinery damage and process downtime due to compressor surge and overspeed can be from thousands to millions of dollars for large continuous chemical or petrochemical plants. This text demonstrates how to select the proper control schemes and instrumentation for centrifugal and axial compressor through-put and surge control. More material is devoted to surge control because surge control is more difficult and the consequences of poor control are more severe.

Special feedback and open-loop backup control schemes and fast-acting instrumentation are needed to prevent surge due to the unusual nature of this phenomenon. In order to appreciate the special instrument requirements, the distinctive characteristics of centrifugal and axial compressors and the surge phenomenon are described. This text focuses on the recent advancements in the description of surge by E.M. Greitzer (Ref. 15). Simple electrical analogies are used to reinforce the explanation. Simulation program plots using the Greitzer model of surge are used to graphically illustrate the oscillations of pressure and flow that accompany different degrees of severity of surge. Extensive mathematical analysis is avoided. A few simple algebraic equations are presented to help quantify results, but the understanding of such equations is not essential to the selection of the proper control schemes and instrumentation.

The surge feedback control scheme is built around the type of controller set point used. In order to appreciate the advantages of various set points, the relationship of the location of the set point relative to the surge curve and the effect of operating conditions on the shape and location of the surge curve are described. The need for and the design of an open-loop backup scheme in addition to the feedback control scheme are emphasized. The use and integration of process and manual override control schemes without jeopardizing surge protection are also illustrated.

Many of the transmitters, digital controllers, and control valves in use at this writing are not fast enough to prevent surge. This text graphically illustrates the effect of transmitter speed of response on surge detection. The effects of transmitter speed of response, digital controller sample time, and control valve stroking time on the ability of the control scheme to prevent surge are qualitatively and quantitatively described. The modifications of control valve accessories necessary for fast throttling and the maintenance requirements are detailed.

The interaction between throughput and surge control and multiple compressors in parallel or series can be severe enough to render the surge control scheme ineffective or even to drive a compressor into surge. This text describes the detuning and decoupling methods used to reduce interaction.

The computational flexibility and power of modern computers facilitates on-line monitoring of changes in compressor performance and its surge curve. This text describes how computers can be used to predict impending compressor damage and the extent of existing damage by vibration frequency analysis. It also describes how computers can be used to gather pressure and flow measurement data to update the surge curve on a CRT screen.

1-2 AUDIENCE AND PREREQUISITES

This text is directed principally to the instrumentation and process control engineers who design or maintain compressor control systems. Process, mechanical, startup, and sales engineers can also benefit from the perspective gained on the unusual problem of compressor surge and the associated need for special instrumentation. Since instrument maintenance groups are genuinely concerned about the proliferation of different types and models of instrumentation, it is critical that the project team members be familiar enough with surge control to be able to justify the use of special instrumentation. Process and mechanical engineers also need to learn how the compressor dimensions and operating conditions can make surge control more difficult and how the piping design can make surge oscillations more severe.

In order for the reader to understand the physical nature of surge, it is desirable that he or she be familiar with some of the elementary principles of gas flow. The reader should know that a pressure difference is the driving force for gas flow, that gas flow increases with the square root of the pressure drop until critical flow is reached, and that gas pressure in a volume will increase if the mass flow into the volume exceeds the mass flow out of the volume and vice versa. If the reader is also comfortable working with algebraic equations and understands such terms as molecular weight, specific heat, efficiency, and the speed of sound, he or she can use the equations presented to describe the surge oscillations and the surge curve. However, the assimilation of these equations is not essential to understanding the control problem and the control system requirements.

In order for the reader to understand the control schemes and special instrumentation requirements, it is desirable that he or she be familiar with the structure, terminology, and typical instrument hardware for a pressure and flow control loop. Specifically he or she should know the functional relationship between the controller, the control valve, and the transmitter; know the terms remote-local set point, feedback control, automatic-manual operation, proportional (gain) mode, integral (reset) mode, and sample time; and know the physical differences between diaphragm and piston actuators, rotary and globe control valves, positioners and boosters, and venturi tubes and orifice plates. The structure, terminology, and hardware for ratio control, override, and decoupling are described in the text as the application is developed.

1-3 LEARNING OBJECTIVES

The surge and throughput control loops for a single compressor appear deceptively simple. However, the success of these loops depends upon the engineer's attention to many details, each of which are critically important. These loops are typically protecting a large capital investment in machinery, protecting against a loss of production due to machinery repair or replacement, and determining the efficiency of a large energy user. The overall goal of this text is to instruct the reader on how to properly design and maintain compressor control loops. The specific individual goals necessary to achieve the overall goal are:

- Learn how the compressor characteristics and operating conditions affect the potential for surge, the surge curve, and the surge controller set point.

- Learn how the compressor and piping design affect the frequency and amplitude of the flow and pressure oscillations during surge.
- Learn how the surge cycles cause compressor damage.
- Learn the relative advantages of different types of instruments in detecting surge.
- Learn how fast the approach to surge will be and how fast the flow reversal is at the start of the surge cycle.
- Learn how to generate the surge curve and the set point for the surge controller for different compressors and operating conditions.
- Learn why a backup open loop is needed in addition to the feedback loop for surge control and how to design one.
- Learn how fast the transmitter, the controller, and the control valve must be to prevent surge.
- Learn how to modify and maintain the control valve accessories to meet the stroking speed requirement.
- Learn how the surge and throughput control system designs affect the operating efficiency of the compressor.
- Learn how to assess the severity of interaction between the surge and throughput control loops and how to reduce it.
- Learn how to use a computer to monitor changes in compressor performance for maintenance and advisory control.

1-4 DEFINITION OF TERMS

We frequently take for granted that others understand the terms we use in the special areas of instrumentation and control. However, misunderstandings or incorrect interpretations can become a major obstacle to learning the concepts. To avoid this problem, the definitions of important terms are summarized below.

axial compressor—A dynamic compressor whose internal flow is in the axial direction.

centrifugal compressor—A dynamic compressor whose internal flow is in the radial direction.

compressor characteristic curve—The plot of discharge pressure versus suction volumetric flow for a typical compressor speed or vane position at a specified suction temperature, pressure, and molecular weight. A family of curves is depicted for variable speed or variable vane position compressors.

compressor diffuser—The stationary passage around the compressor impeller where a portion of the velocity pressure is converted to static pressure.

compressor (dynamic)—A compressor that increases the pressure of a gas by first imparting a velocity pressure by rotating blades and then converting it to a static pressure by a diffuser. Dynamic compressors are either centrifugal or axial. If the discharge pressure is less than 10 psig, dynamic compressors are usually called blowers. If the discharge pressure is less than 2 psig, dynamic compressors are usually called fans.

compressor guide vane—Stationary blades at the inlet eye of the impeller that direct the angle of the gas flow into the impeller. The angle of the blades can be adjustable to impart varying amounts of rotation to the gas. This angle is with or against the rotation imparted by the impeller. The adjustable angle varies the capacity and the discharge pressure of the impeller.

compressor impeller—The blades on the rotating compressor shaft that impart the velocity to the entering gas.

compressor map—the compressor characteristic curves and the surge curve at a specified suction temperature, pressure, and molecular weight.

compressor rotor—The rotating element in the computer that includes the compressor impeller and shaft.

compressor stage—each set of compressor blades plus diffuser is a compressor stage. There can be multiple stages within the same housing or there can be a single housing for each stage with a heat exchanger in between.

compressor stall—Unstable flow pattern in a compressor where the forward flow stops in localized regions around the impeller.

compressor stall or surge curve—The curve drawn through the point of zero slope on each compressor characteristic curve. If the operating point is to the right of this curve, compressor operation is stable. If the operating point is to the left of this curve, compressor surge or stall can occur.

compressor surge—Unstable flow pattern in a compressor where the total flow around the impeller alternately stops or flows backwards and then flows forward.

compressor thrust—The axial displacement of the compressor shaft that can occur during surge.

compressor vibration—the radial oscillation of the compressor shaft that can occur during surge.

controller gain—The mode that changes the controller output by an amount equal to the change in error multiplied by the controller gain. The proportional band is the percent change in error necessary to cause a full-scale change in controller output. Proportional band is the inverse of controller gain multiplied by 100.

controller rate—The mode that changes the controller output by an amount proportional to the derivative of the error. The derivative time is that time required for the proportional band contribution to equal the derivative (rate) mode contribution for a ramp error.

controller reset—The mode that changes the controller output by an amount proportional to the integral of the error. The integral time is that time required for the integral (reset) mode to equal (repeat) the proportional band contribution for a constant error. Most controllers use the inverse of integral time so that the reset setting units are repeats per minute.

controller reset windup—The condition of controller output when the reset contribution to the controller output exceeds the output change of the controller. The controller output is at the upper or lower extremity of its range and will not change until the measurement crosses set point. Most anti-reset windup options for controllers limit the reset contribution so that it plus the proportional band contribution does not exceed an adjustable upper and lower output limit.

steady-state gain—The final change in output divided by the change in input (all the oscillations have died out). It is the slope of the plot of the steady-state response versus input. If the plot is a straight line, the gain is linear (slope is constant). If the plot is a curve, the gain is nonlinear (slope varies with operating point).

steady-state response—The final value of an output for a given input (all oscillations have died out). The compressor characteristic curve is a plot of the steady-state response of compressor discharge pressure for a given suction flow.

time constant—The time required for the output to reach 63 percent of its final value with an exponentially decreasing slope.

time delay dead time—The time required for an output to start to change after an input change.

transient response—The value of an output as it varies with time after an input change. The oscillations of compressor discharge pressure and suction flow during surge are transient responses.

valve positioner—A proportional-only pneumatic controller mounted on a control valve whose measurement is valve position and whose set point is the output of a process controller or manual loader (via an I/P transducer for an electronic loop). The gain of this position feedback controller is typically greater than 100 (the proportional band is less than 1 percent).

volume booster—A pneumatic relay (usually 1:1—the change in output signal is equal to the change in input signal) that has a much greater air flow capacity than positioners or I/P transducers. The greater air flow capacity increases the speed of the control valve stroke. How fast the pressure in the actuator volume tracks the pneumatic signal depends on the supply and exhaust flow capacity of the booster and the size of the actuator.

Important new terms such as *surge* and *stall* will be defined in greater detail in subsequent sections. The first time an important new term is introduced, it will be italicized for reference and emphasis.

Description of Compressors

2-1 GENERAL TYPES

The two general types of compressors are positive displacement and dynamic. The positive displacement compressor increases the gas pressure by confinement within a closed space. *Reciprocating compressors* are positive displacement compressors where the closed space in a cylinder is decreased by a piston to compress the gas.

Figure 2-1 shows the steps in the compression cycle of a reciprocating compressor. In step 1 the piston is completely withdrawn from the cylinder. Since the suction pressure is greater than the pressure in the empty cylinder, the inlet valve opens and the gas fills the cylinder until the cylinder pressure equals the suction pressure. In step 2 the piston has partially stroked and reached a position where the cylinder pressure is greater than the suction pressure but less than the discharge pressure. Both the inlet and outlet valves are closed so that there is no suction or discharge flow of the gas. In step 3 the piston has fully stroked and compressed the gas enough to cause the cylinder pressure to exceed the discharge pressure. The outlet valve opens and the gas flows out of the chamber. Notice that at the fully stroked position of the piston there is a clearance volume. If this clearance volume is increased, the capacity of the compressor is decreased. The capacity of reciprocating compressors can also be decreased by decreasing the number of cylinders in service or the speed of the cycle. The capacity cannot be decreased

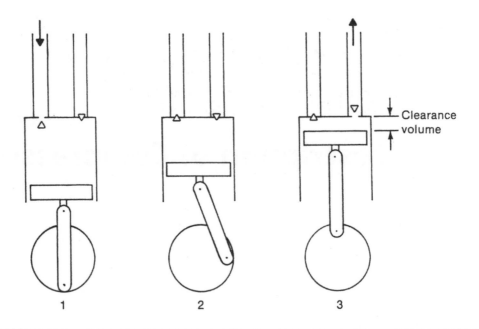

Figure 2-1   Compression Cycle of a Reciprocating Compressor

by throttling the discharge flow or the suction flow with a control valve. The suction volumetric flow at suction conditions (actual cubic feet per minute or acfm) for a reciprocating compressor is essentially independent of gas composition or gas pressure within its design limits. Relief valves are installed on the discharge of reciprocating compressors because the discharge pressure can rise and exceed the pressure rating of downstream equipment if a downstream valve is closed.

Figure 2-2 shows the *compressor characteristic curve* for a reciprocating compressor, which is a plot of pressure rise versus inlet volumetric flow at a single speed. Also shown is the *demand load curve* for the system, which is a plot of the total pressure drop due to downstream equipment, piping, and valves versus inlet volumetric flow. The pressure drop of the demand load curve increases with approximately the square of the flow. Notice that as the resistance of the system increases (the pressure drop at a given flow increases due to throttling or fouling), the load curve shifts upward and intersects the characteristic curve at L2 instead of L1. The intersection of the load curve and the characteristic curve denotes the *operating point* of the system since the discharge pressure equals the pressure drop at this point. The change in inlet flow from operating point L1 to L2 is small

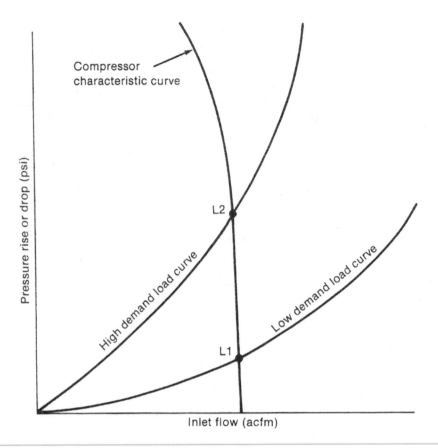

Figure 2-2   Reciprocating Compressor Characteristic Curve

because the characteristic curve is nearly vertical for a reciprocating compressor. The characteristic curve is not perfectly vertical because the volumetric efficiency decreases with pressure.

Reciprocating compressors can maintain a constant flow even though the operating conditions of the system may change. However, discharge pressure and flow will be pulsating at the frequency of the cycle of the reciprocating compressor. The discharge volume of the system is frequently large enough to dampen the amplitude of these high-frequency pulsations. In some high-pressure applications, the gas pressure pulsation is sufficient to cause cracks in valves, piping, and cooler tubes. The installation of a dead-ended volume as a tee in the pipeline can dampen these oscillations if the volume and distance from the compressor discharge is properly selected based on the pulsation frequency (Ref. 11).

Dynamic compressors increase gas pressure by first accelerating the gas and then converting the increased velocity energy of the gas into pressure. The gas is accelerated by rotating blades called an *impeller* on a shaft driven by a motor or a turbine. The velocity energy of the gas is converted to pressure in a stationary passage of increasing cross-sectional area called the *diffuser*. Dynamic compressors are usually called blowers, if the discharge pressure is less than 10 psig, and fans, if the discharge pressure is less than 2 psig (Ref. 22). At the time of this writing the boundaries between the classifications of compressor, blower, and fan are not distinct. In general the machines used to supply gas to furnaces, dryers, and HVAC systems are called blowers or fans. The discharge pressure of these machines are usually measured in inches of water column instead of psig. Dynamic compressors, fans, and blowers are either centrifugal or axial.

Figure 2-3 shows the approximate operating ranges of the three major classes of compressors. The centrifugal and axial compressor classes are extended to show the ranges of blowers and fans. Since the ranges overlap and multiple compressors can be used to extend a range, criteria other than operating range are also important to determine what type of compressor should be selected for a given application. Some of these criteria are cost, maintenance, efficiency, erosion, fouling, and corrosion. Reciprocating compressors generally cost more than centrifugal compressors for equivalent capacities. Reciprocating compressors require more mechanical maintenance due to vibration, gas pulsation, cylinder valve flutter, and cylinder lubrication (Ref. 11). Reciprocating compressor efficiency varies from 0.75 to 0.85. Centrifugal compressor efficiency increases with flow from about 0.65 at 1000 acfm to 0.77 at 200,000 acfm. Axial compressor efficiency increases with flow from about 0.81 at 70,000 acfm to 0.83 at 600,000 acfm (Ref. 10). Centrifugal compressor construction can be more easily modified for erosive, dirty, or corrosive gases. Centrifugal impellers are constructed with alloys, platings, or coatings for abrasion and corrosion resistance or are constructed to be self-cleaning by centrifugal force or scraping action (Ref. 26). Axial compressors are used primarily for air or clean noncorrosive gases.

Key Concepts

- Compressors are either positive displacement or dynamic.
- Positive displacement compressors increase gas pressure by confinement.
- Dynamic compressors increase gas pressure by acceleration.

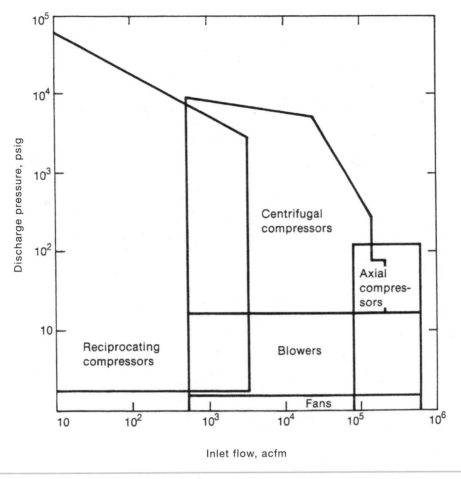

Figure 2-3   Approximate Ranges of Compressors

- The characteristic curve is a plot of pressure rise versus flow.
- The load curve is a plot of system pressure drop versus flow.
- Positive displacement compressors have a vertical characteristic curve.
- Positive displacement compressors require more mechanical maintenance.
- Dynamic compressors are either centrifugal or axial.
- Centrifugal compressors can handle dirty, corrosive, and erosive gases.
- Axial compressors are more efficient than centrifugal compressors.

2-2 CENTRIFUGAL COMPRESSORS

Figure 2-4 shows a cross section of a *centrifugal compressor* with multiple impellers. The gas flows radially from the tip of the impeller into the diffuser and then into the next impeller. Each impeller/diffuser set represents one stage of compression in a multistage compressor. Some authors reserve the use of the term "multistage compressor" for those compressors that contain separate housings for the stages and heat exchangers (called intercoolers) between the housings to cool the gas before recompression.

Figure 2-5 shows the compressor characteristic curve for a centrifugal compressor at a single speed and with specific operating conditions. This curve is similar in shape to the head curve for a centrifugal pump. Also shown is the demand load curve. As the resistance of the downstream system increases, the load curve

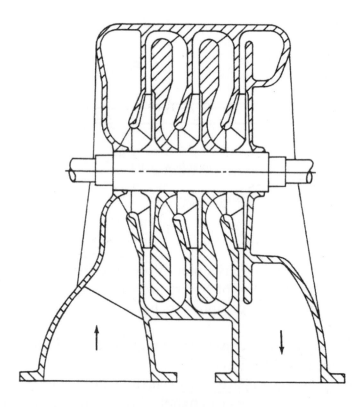

Figure 2-4   Centrifugal Compressor Cross Section
(Courtesy Compressed Air and Gas Institute)

shifts upward to a new position. The intersection point of the load curve and the characteristic curve (operating point) moves from L1 to L2. The volumetric flow decreases appreciably, but the discharge pressure increases only slightly. For a perfect centrifugal compressor without any internal losses, the characteristic curve would be horizontal and the discharge pressure would remain constant. A centrifugal compressor is particularly suited for handling large gas volumes for constant pressure and variable throughput control. However, the position of the characteristic curve, and thus the intersection point with the load curve, depends on the suction operating conditions and speed. Section 4 will discuss in detail the effect of these operating conditions. This movement of the characteristic curve facilitates variable pressure as well as variable throughput control by suction throttling or speed control.

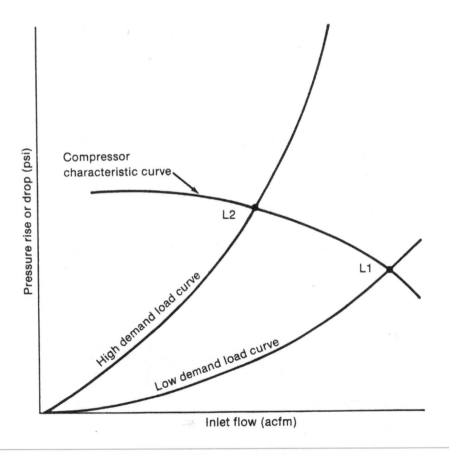

Figure 2-5   Centrifugal Compressor Characteristic Curve

Key Concepts

- The centrifugal compressor flow path is radial.
- The centrifugal compressor characteristic curve is relatively flat.

2-3 AXIAL COMPRESSORS

Figure 2-6 shows a cross section of an axial compressor with multiple impellers. The gas flows axially along the shaft of the compressor from one impeller to another, directed by the stationary vanes. Each impeller/stationary vane set represents one stage of compression. Since an impeller in an axial compressor develops about half the pressure rise of an equivalent impeller in a centrifugal compressor, an axial compressor casing will frequently have twice as many impellers in its housing (Ref. 10).

Figure 2-7 shows the compressor characteristic curve for an axial compressor at a single speed and with specific operating conditions. Also shown is the load curve and its shift upwards that moves the operating point from L1 to L2 as the

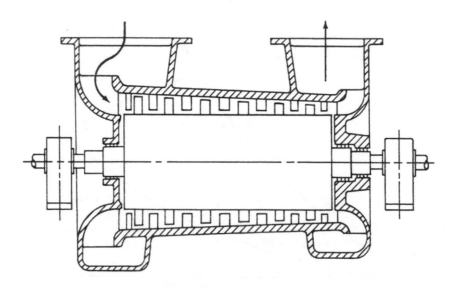

Figure 2-6   Axial Compressor Cross Section
(Courtesy Compressed Air and Gas Institute)

system resistance increases. The pressure increases more than the flow decreases since the characteristic curve is more vertical than horizontal. An axial compressor is particularly suited for constant flow and variable pressure control. However, since the position of characteristic curve depends on suction operating conditions, vane position, and speed, these variables can be manipulated to provide variable throughput as well as variable pressure control.

Key Concepts

- The axial compressor flow path is axial.
- The axial compressor characteristic curve is relatively steep.

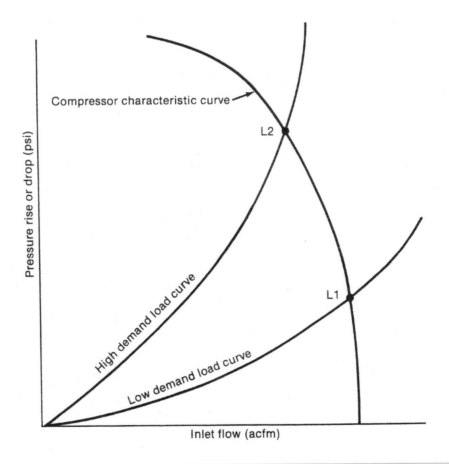

Figure 2-7   Axial Compressor Characteristic Curve

QUESTIONS

1. Why can't the flow through a reciprocating compressor be controlled by throttling a control valve on its discharge?

   ANSWER

2. Why would a centrifugal compressor be preferred rather than an axial compressor at constant speed if a constant discharge pressure is desired?

   ANSWER

3. Why would an axial compressor be preferred rather than a centrifugal compressor for 100,000 acfm of clean air if the cost is the same?

   ANSWER

STUDENT SUMMARY NOTES AND QUESTIONS FOR INSTRUCTOR

Description of Surge

3-1 SURGE VERSUS STALL

Surge is a dynamic instability that occurs in dynamic compressors. Surge can also occur in axial and centrifugal pumps and blowers, but the occurrence is less frequent and the damage less severe. Pump surge seldom occurs except infrequently during cavitation and two-phase flow (Ref. 15). Blower surge does occur but does not damage the blower internals unless the pressure rise exceeds 2 psi and the size exceeds 150 BHP (brake horsepower) (Ref. 22). Surge does not occur in positive displacement compressors and pumps.

Stall is another instability that occurs in dynamic compressors and is sometimes confused with surge. The consequences of stall are usually less severe. Pump stall has caused extensive vibration in the piping of boiler feedwater systems (Ref. 15). Stall will be described in enough detail to distinguish it from surge and to understand how it can develop into surge.

As the flow through a compressor is reduced, a point is reached where the flow pattern becomes unstable. If the flow oscillates in localized regions around the rotor, the instability is called *stall*. For axial compressors these regions of unstable forward flow can extend over just a few blades or up to 180 degrees around the annulus in the compressor. In rotating stall, the region of unstable forward flow rotates around the annulus. The average flow across the annulus is still positive. The frequency of the localized flow oscillations ranges from 50 to

100 hertz. The stall frequency is some fraction of the compressor speed. The stall frequency does not depend on the design of the piping system in which the compressor is installed. Stall can develop into a more global type of instability called *surge*. In surge the average flow across the annulus goes through large amplitude oscillations. The frequency of these oscillations ranges from 0.5 to 10 hertz. The surge frequency depends on the designs of both the compressor and piping system. The surge frequency for most industrial compressor installations is slightly less than 1 hertz.

Figure 3-1 shows a *compressor map* for a variable-speed centrifugal compressor. A compressor map is the single most important piece of information for describing surge. A compressor map shows the compressor characteristic curve for different operating conditions. Each curve traces the rise in discharge pressure developed by the compressor as the suction flow is varied for a given operating condition (such as speed). The operating condition is not limited to that shown by the few curves plotted but is continuously adjustable to intermediate values. The X axis is almost always volumetric flow in acfm at the stated suction temperature, pressure, and molecular weight. The Y axis is either the discharge pressure in psia, the ratio of discharge pressure to suction pressure (dimensionless), or the differential pressure rise from suction to discharge in psi. The compressor characteristic curves supplied by the compressor manufacturer cover the negative sloped region and end where the slope approaches zero. (A negative slope means the pressure will change in the opposite direction for a change in flow, and a zero slope means the pressure will not change at all for a change in flow.) If a curve is drawn through the point of zero slope for each characteristic curve, the region to the left of this line is where the instabilities of surge and stall occur. In the literature, this line is called either the stall line or the surge curve. In this text it will be always called the surge curve. The characteristic curve to the left of this curve is difficult to obtain, but its general shape resembles that shown in Figure 3-1. The negative flow portion of the curve can be found by supplying pressurized gas to the compressor discharge to force a steady negative flow. The positive sloped portion of the curve is obtained by smoothly connecting a third-order polynomial curve between the negative sloped negative flow and the positive flow portions of the curve.

Surge can be better understood by visualizing a block or throttle valve closing downstream of the compressor. As this valve closes, the suction flow decreases, and the operating point moves to the left along a characteristic curve. If the compressor speed or the guide vane position does not change, the operating point eventually passes by the point of zero slope on the characteristic curve. Just to

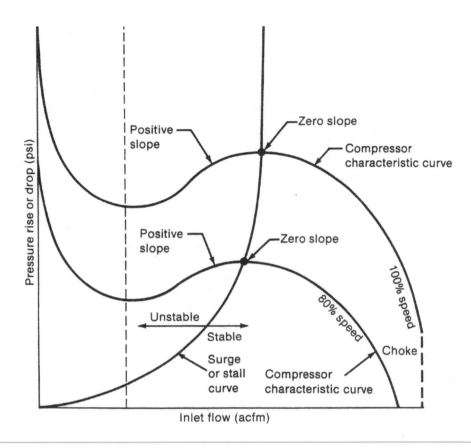

Figure 3-1   Compressor Map

the left of this point, which is the intersection of the surge curve with the characteristic curve, the pressure developed by the compressor is less than the pressure in the piping between the compressor discharge and the closing valve. The forward flow through the compressor stops and reverses direction. The gas flows from the discharge piping back through the compressor to the suction piping. The pressure in the discharge piping starts to drop. When this pressure is below the pressure developed by the compressor, forward flow starts again. If the gas in the discharge piping is still trapped by the closing block or throttle valve, the gas pressure builds up and the surge cycle is repeated. Note that the alternating pressure buildup and decay cycle does not require the flow to actually reverse direction but only to alternately decrease below or increase above the flow rate through the block or throttle valve. Only in severe surges does the flow through the compressor actually reverse direction.

Whether a compressor system is stable or unstable depends on how the operating point reacts to a disturbance. If the operating point returns to its initial value after a disturbance, the system is stable. If the deviation of the operating point from its initial value grows or continually oscillates after a disturbance, the system is unstable. A uniform growing deviation is classified as a static instability and an oscillating deviation of growing or constant amplitude is classified as a dynamic instability. Surge is a dynamic instability.

Key Concepts

- Surge and stall occur in dynamic compressors.
- Stall consists of localized flow oscillations around the rotor.
- Surge consists of total flow oscillations around the whole rotor.
- The surge curve is the zero slope points of the characteristic curves.

3-2 STATIC INSTABILITY

Static instability is a uniform (non-oscillatory) growing deviation. Static instability will occur when the slope of the compressor characteristic curve is greater than the slope of the demand load curve.

As previously mentioned, the demand load curve is a plot of how the pressure drop in the system (due to piping, equipment, and valve resistance) increases with flow. The general criterion is:

For static instability,

$$S_c > S_l \quad (3\text{-}1)$$

where:

S_c = slope of the compressor characteristic curve (psi/acfm)
S_l = slope of the demand load curve (psi/acfm)

Figure 3-2 illustrates how the criterion for static instability can be checked by drawing the demand load curve on a compressor map. The compressor map shows stability at operating point 1 but static instability at operating point 2 where the compressor curve slope is steeper than the load curve slope. Since the compressor curve and the load curve are both steady-state curves, this instability criterion can

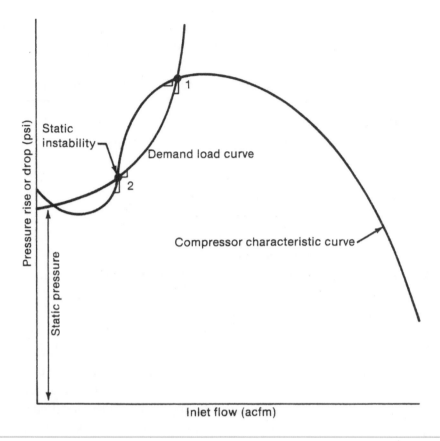

Figure 3-2   Static Instability Criterion

be checked at one steady state or a succession of steady states without knowledge of the system dynamics; hence the name, static instability. An electrical circuit analogy consists of a voltage or current source and a resistance. In this analogy, voltage represents pressure, current represents flow, and electrical resistance represents piping, equipment, and valve resistance. If the compressor characteristic curve is nearly horizontal, which is typical for centrifugal compressors, the circuit has a nonideal voltage source. If the compressor characteristic curve is nearly vertical, which is typical for axial compressors, the circuit has a nonideal current source. This circuit analogy has no dynamic components such as inductors or capacitors so the current flow does not vary with time if the system is stable. The circuit develops static instability if the slope of the source curve is greater than the slope of the resistance curve for voltage versus current. If the slope of the resistance is always positive (voltage drop always increases as current increases) and the slope of the source curve is always negative (voltage rise always decreases as current increases), the circuit is stable. If the slope of the source curve is positive and greater than the slope of the resistance curve, the following sequence of events occurs for any disturbance that causes an increase in voltage output:

(1) Increase in source voltage
(2) Increase in voltage (driving force) across the resistance
(3) Increase in current
(4) Increase in source voltage (sequence repeats)

The corresponding sequence for a compressor system is:

(1) Increase in compressor pressure
(2) Increase in pressure (driving force) across the load
(3) Increase in flow
(4) Increase in compressor pressure (sequence repeats)

A similar sequence and exponential growth of any positive deviation occurs in the open-loop response of temperature for runaway exothermic reactors and in the open-loop response of cell concentration for runaway biological reactors (Ref. 21). In a compressor system, static instability occurs when the load curve is nearly flat and intersects the negative sloped portion of the compressor curve. The load curve is nearly flat if a piece of equipment with pressure control such as a reactor, condensor, or absorber is connected to the discharge of a compressor by a short duct or pipe without a throttle valve. Figure 3-2 illustrates this case. The pressure at the starting point of the load curve at zero flow is the static pressure or set point of the pressure control loop for the downstream equipment. The slope of the load curve is proportional to the flow multiplied by twice the resistance of the duct or pipe. Since this duct or pipe is short, its resistance is low, and the slope of the load curve is low at low flow.

$$S_l = 2 \cdot K_l \cdot Q \quad (3\text{-}2)$$

where:

K_l = resistance of load (psi/(acfm · acfm))
Q = volumetric flow at suction conditions (acfm)
S_l = slope of the demand load curve (psi/acfm)

If the duct or pipe is long, the resistance coefficient K_l can still be small if the pipe or duct diameter is large with respect to the flow capacity of an individual compressor. This situation occurs when multiple parallel compressors are supplying a distribution header. The header diameter is selected based on the total flow capacity of all the compressors. The change in pressure drop in the header with a change in flow from an individual compressor is small, and thus the slope of the load curve on the compressor map is small.

Static instability occurs only if the compressor system can reach an operating point on the positive sloped portion of the compressor characteristic curve. Normally the flow drops precipitously when the operating point reaches the zero sloped portion of the compressor curve and surge oscillations develop, which is a dynamic instability.

Key Concepts

- Static instability is more likely to occur for parallel compressors.
- Dynamic instability usually occurs before static instability.

3-3 DYNAMIC INSTABILITY

Dynamic instability is characterized by growing oscillations. Dynamic instability can occur if either the discharge plenum volume or the compressor impeller speed is large enough to cause a system response parameter to exceed a minimum value and the slope of the compressor characteristic curve is positive. The general criterion is:

For dynamic instability,

$$B > B_m \quad (3\text{-}3)$$

$$S_c > 0 \quad (3\text{-}4)$$

$$B = \{N \cdot \sqrt{V_p/(A_c \cdot L_c)}\} / (2 \cdot a) \quad (3\text{-}5)$$

where:

a = speed of sound in the gas (ft/sec)

A_c = cross-sectional area of flow path in compressor (sq ft)

B = system dynamic response parameter (dimensionless)

B_m = minimum B for dynamic instability (typically 0.1 to 1.0)

L_c = length of flow path in compressor (ft)

N = compressor impeller speed (rev/sec)

V_p = volume of the plenum (see Figure 3-3) (cu ft)

Figure 3-3 shows a simple compressor system with the components identified for the dynamic instability criterion. The plenum is any enclosed volume of gas. In a compressor installation, the plenum may be a vessel or distribution header between the compressor and the block or throttle valve. If the valve is installed directly at the end of a section of pipe or duct connected to the compressor discharge, the volume of this section of pipe or duct can be used as an equivalent plenum volume without seriously degrading the accuracy of Equation 3-3. The compressor flow path length and area are small enough and the plenum volume is large enough in most industrial installations to satisfy the criterion for dynamic instability when the operating point crosses over from the negative to the positive sloped portion of the compressor characteristic curve. However, the oscillation amplitude stops growing due to the nonidealities and nonlinearities of compression in industrial applications. The resulting constant amplitude oscillation

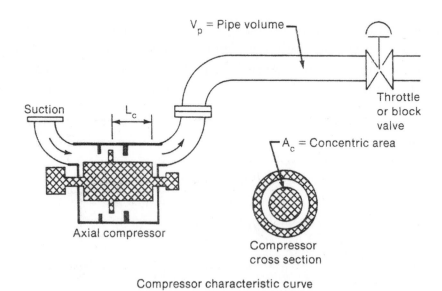

Figure 3-3   Simple Compressor System

is called a limit cycle. Surge is a dynamic instability that develops into a limit cycle. Stall is not a dynamic instability because the amplitude of the oscillations decreases (decays) with time. Thus the minimum B value (typically 0.1 to 1.0) is the boundary line between surge and stall. Surge occurs when B is above the minimum B, and stall occurs when B is below the minimum B. Also, the severity of surge is proportional to the magnitude of B. Note that this minimum B parameter depends on the compressor speed and the system dimensions and is independent of compressor manufacturer and model number. Thus, different compressors can be compared on a consistent basis.

Key Concepts

- Surge is a dynamic instability (sustained oscillations).
- The severity of surge increases with speed for a given installation.
- The severity of surge increases with volume for a given installation.

3-4 CHARACTERISTICS OF SURGE

Deep surge starts with a precipitous drop in flow. The flow will typically drop from its set point to its minimum (possibly negative) in less than 0.05 second. No other physical phenomenon can cause such a drop in flow.

Figure 3-4 shows the precipitous drop in flow measured by both a slow pneumatic transmitter and a fast electronic d/p flow transmitter. Since this precipitous drop in flow is unique to surge, it can be used as a trigger to actuate an interlock that will open the surge valves or start a surge counter. This drop can be detected by taking the derivative of the d/p signal if the flow measurement signal is not noisy and the transmitter is fast enough (see Section 6-4 for more detail on surge detection methods and Section 7-4 for the effect of transmitter time constant on surge detection).

The period of the surge oscillation (time between successive peaks in flow) is shorter than the control loop period unless fast instruments are used and the controller is tuned tight. The surge period is typically less than 2 seconds, while the loop period is usually greater than 4 seconds. Since the surge period is shorter than the loop period, by the time the control loop reacts to a particular portion of the surge cycle, that portion of the surge cycle is long gone. The corrective action by the control loop may become in phase with subsequent surge

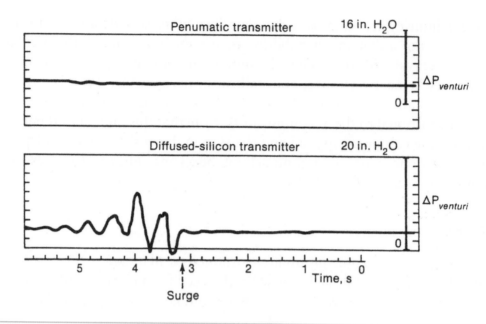

Figure 3-4   Precipitous Drop in Flow Measured by Two Transmitters
(Courtesy Compressor Controls Corporation)

cycles and can accentuate the severity of the surge. If the surge period is less than one-fourth the loop period, the surge cycle can be considered as equivalent to uncontrollable noise (Ref. 21). The minimum loop period is approximately equal to 4 times the summation of the instrument time constants and dead times. (See Section 12-1 for more details on how to estimate the loop period for tuning the surge controller.) The minimum surge period that occurs when B is about equal to the minimum B for dynamic stability can be estimated by the following equation:

$$T_s = \{2\pi \cdot \sqrt{(L_c \cdot V_p)/A_v}\}/a \quad (3\text{-}6)$$

where:

    a = speed of sound (ft/sec)

    A_c = flow path cross-sectional area in compressor (sq ft)

    L_c = flow path length in compressor (ft)

    T_s = surge oscillation period (sec)

    V_p = volume of the plenum (cu ft)

Equation 3-6 shows that the surge oscillation period increases as the plenum volume increases. Most industrial compressor installations have a check valve installed on the compressor discharge to prevent backflow during the opening of the surge valves or during the flow reversal caused by surge. This check valve is particularly important for fluidized bed reactor applications to prevent the backflow of catalyst or flammable mixtures. It is also important for parallel compressor applications to prevent pumping gas from one compressor back through another. The check valve will slam shut during the start of surge and will effectively reduce the plenum volume to that of the piping between the compressor and the check valve. Thus, surge oscillation periods are not large even when there is a long header or vessel between the compressor and the block or throttle valve.

The plenum pressure does not precipitously drop as does suction flow at the start of surge. Also the oscillation amplitude in pressure is usually less than that in flow during surge. Therefore surge counters or backup control systems that try to detect surge by pressure measurement are liable to experience false or missed surge counts or trips. This is particularly true if pneumatic pressure transmitters are used or if the pressure is noisy or swings with load.

The flow oscillation peaks correspond to the filling of the plenum, and the valleys correspond to the emptying of the plenum. The time duration of the peaks and the valleys is typically much longer than the time duration of the transition between the peaks and valleys. The time duration of the peaks and valleys depends on the piping system friction (resistance) and volume (capacitance), while the time duration of the transition depends on the fluid inertia (inductance). The system dynamic response parameter B is proportional to the ratio of the pressure driving force to the inertial impedance. B is representative of the capability to accelerate the gas. As parameter B increases, the amplitude of the oscillations becomes larger and the shape becomes more non-sinusoidal.

Figure 3-5a shows the path traced out by the operating point on a compressor map for severe surge where B is about 6 times the minimum B for dynamic stability as a throttle valve is closed downstream of the compressor. Figure 3-5b shows the oscillations in suction flow and discharge pressure that correspond to the path traced out by the operating point in Figure 3-5a. Notice that the oscillation amplitude is extremely large and the shape is non-sinusoidal. In Figure 3-5a, the operating point starts at point A and moves to the left along the compressor characteristic curve as the throttle valve closes. When the operating point reaches point B, which is where the compressor characteristic curve slope is zero, the operating point jumps to point C. This jump corresponds to the precipitous drop in flow that signals the start of the surge cycle. The operating point

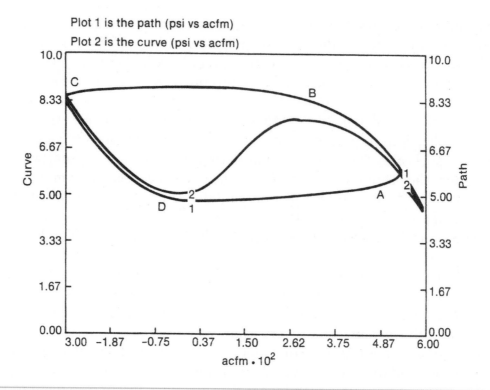

Plot 1 is the path (psi vs acfm)
Plot 2 is the curve (psi vs acfm)

Figure 3-5a   Operating Point Path and Compressor Curve (Severe Surge)

cannot follow the positive sloped portion of the compressor curve because the reduced flow out of the plenum, caused by the closing throttle valve, requires that the plenum pressure increase instead of decrease per the curve. (The plenum pressure must increase if the flow into the plenum exceeds the flow out of the plenum.) The flow pattern around the impeller breaks up, and the flow reverses direction to backward flow from the discharge volume to the compressor suction. After this jump to point C, the operating point follows the compressor curve from point C to point D as the plenum volume is emptied due to reverse flow. When the operating point reaches point D, which is where the compressor characteristic slope is zero again, the operating point jumps to point A. The operating point cannot follow the positive sloped portion of the compressor curve because the reduced forward flow into the plenum volume requires that the plenum pressure decrease instead of increase per the curve. (The plenum pressure must decrease if the flow out of the plenum exceeds the flow into the plenum.) The flow pattern around the impeller is established and the flow forward rapidly increases. After the jump to

Plot 1 is the flow (acfm vs time)

Plot 2 is the pressure (psi vs time)

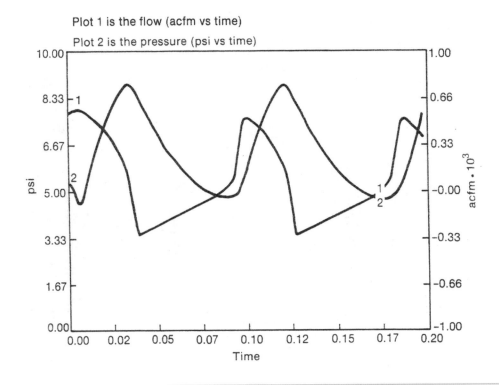

Figure 3-5b   Suction Flow and Discharge Pressure Oscillations (Severe Surge)

point A, the operating point follows the compressor curve to point B as the plenum volume is filled. The surge cycle repeats itself unless the throttle valve or a surge control valve is opened. The operating point follows the compressor curve only during the peaks and valleys of the surge flow oscillations. The jumps on the compressor map correspond to the rapid transition between the peaks and valleys. The oscillation period is about 1.5 times the period predicted by Equation 3-6.

Figure 3-6a shows the path traced out by the operating point on a compressor map for the transition between surge and stall. B has been decreased until it is about equal to the minimum B for dynamic stability by decreasing the speed as a throttle valve is closed downstream of the compressor. Figure 3-6b shows the oscillations in suction flow and discharge pressure that correspond to the path traced out by the operating point in Figure 3-6a. Notice that the oscillation amplitude is an order of magnitude less than that in Figure 3-5b, and the shape is nearly sinusoidal. The B value is actually slightly less than the minimum B because the oscillation amplitude is gradually decreasing, and the path traced out by the

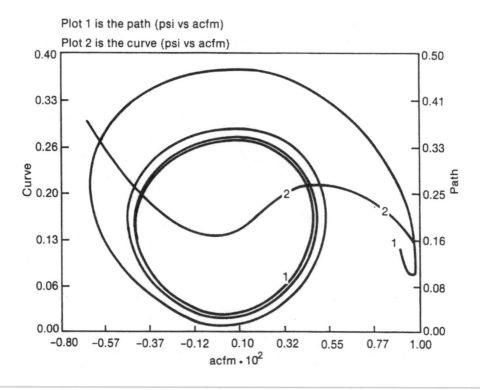

Figure 3-6a   Operating Point Path and Compressor Curve (Surge to Stall Transition)

operating point in Figure 3-6a is gradually spiralling inward. The surge oscillation period is about equal to that predicted by Equation 3-6.

Figure 3-7a shows the path traced out by the operating point on a compressor map for stall where *B* has been decreased to about 0.01 times the minimum *B* for dynamic stability by decreasing the speed as a throttle valve is closed downstream of the compressor. Figure 3-7b shows the oscillations in suction flow and discharge pressure that correspond to the path traced out by the operating point in Figure 3-7a. Notice that the oscillation amplitude is 2 orders of magnitude less than in Figure 3-6b and decays to zero. The path traced out by the operating point in Figure 3-7a rapidly spirals inward and converges to the point of zero flow. The operating point approaches zero flow because the throttle valve downstream closed completely, which stopped flow out of the system except for a small leakage flow. If the throttle valve had remained slightly open, the spiral would have converged to a point of positive flow on the positive sloped portion of the compressor curve. The stall oscillation period is about 1/2 the period predicted by

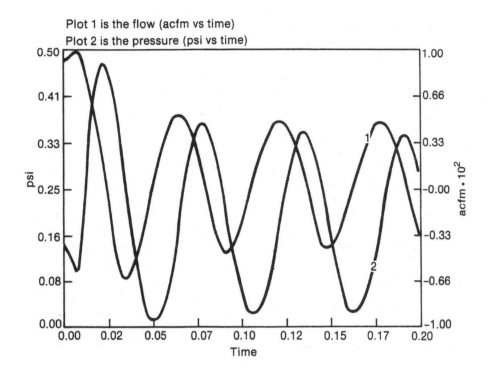

Figure 3-6b   Suction Flow and Discharge Pressure Oscillations (Surge to Stall Transition)

Equation 3-6. This characteristic of compressor response, where the oscillation period decreases as the amplitude decreases, is the opposite of the characteristic of general control system response. The ultimate period of a control system occurs at the transition from stable to unstable operation (similar to the transition from stall to surge). However, for unstable operation (growing oscillations) the control loop period decreases, and for stable operation (decaying oscillations) the loop period increases.

The plots in Figures 3-5 and 3-6 were generated by an advanced continuous simulation language (ACSL) program documented in Appendix B. The program is based on the Greitzer model of surge that was described and experimentally tested by K.E. Hansen et al. (Ref. 16). The ACSL program integrates the differential equations for the momentum and mass balances and computes the stability parameter B, the surge period, and the system flows and pressures. The user must enter data on compressor curve shape, compressor speed, system geometry, surge valve, and throttle valve.

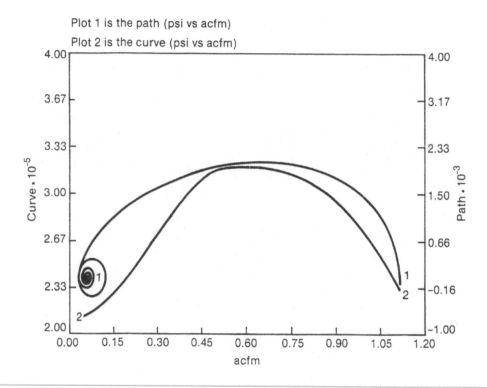

Figure 3-7a   Operating Point Path and Compressor Curve (Stall)

Key Concepts

- Deep surge starts with a precipitous drop in total flow.
- The surge period is much shorter than the control loop period.
- The surge period increases with volume for a given installation.
- The surge period increases slightly with surge severity.
- The surge shape becomes more non-sinusoidal with surge severity.

3-5 CONSEQUENCES OF SURGE

The rapid flow reversals cause extensive radial vibration and axial thrust displacement. The reheating of the same mass of gas during each surge cycle causes a large temperature increase. The gas temperature rise is particularly dramatic for axial compressors, with an observed increase in temperature of 3000°F after 10 surge

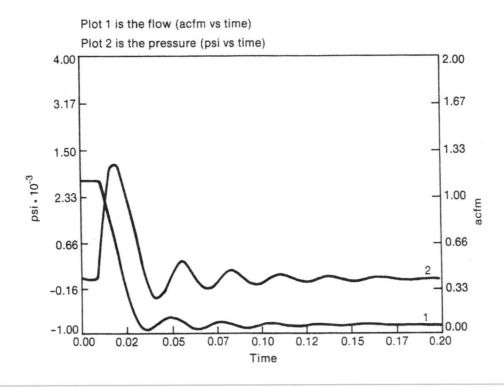

Plot 1 is the flow (acfm vs time)

Plot 2 is the pressure (psi vs time)

Figure 3-7b   Suction Flow and Discharge Pressure Oscillations (Stall)

cycles for one installation (Ref. 26). Vibration and thrust monitors have built-in electronic delays to prevent false alarms or trips due to noise. Temperature sensors have thermal lags due to the resistance to heat transfer of the sensor and thermowell construction (see Section 7). As a result these instruments will usually actuate an alarm or shutdown after several surge cycles have occurred. In the interim, the increase in vibration, thrust, and temperature can cause extensive damage to the compressor. The repair costs can range from thousands to millions of dollars. If the compressor is the sole supplier of gas for a plant unit, the business interruption loss can be much greater than the repair costs, particularly if a replacement rotor must be manufactured, the plant product was sold out at the time of the interruption, or customers switch permanently to a competitor. Even if the damage is not noticeable, the surge can change the internal clearances enough to decrease the efficiency of the compressor. Up to a 0.5 percent loss in efficiency may result from a few cycles of surge (Ref. 26). Repeated surges will cause a significant accumulated loss in efficiency.

The rapid unloading of the impeller during flow breakdown can cause over-speed of the compressor. Field measurements (with an oscillographic recorder) of the speed of one compressor during the start of surge indicated that the compressor acceleration increased to 2000 rpm per sec until power was removed. This increase in the derivative of the speed (acceleration) is characteristic of a runaway or positive feedback open-loop response (Ref. 21). The runaway response was so rapid that the speed controller for the tested compressor could not react and prevent overspeed damage at the start of surge. Consequently, the only alternative was to shut down the compressor when the output of a derivative module indicated the start of a runaway speed response (700 rpm per second). The same module started the oscillographic recorder at 300 rpm per second acceleration to capture a record of the operating conditions just prior to the shutdown. The power was removed from the compressor by closing the steam turbine and expander supply valves in less than one half second.

The flow reversals during surge can produce a booming noise loud enough to make surge a memorable experience for personnel in the vicinity of the compressor. The booming noise can originate from collapsing gas voids within the compressor, the flexing of the suction filter walls, or the slamming shut of the check valve.

KEY CONCEPTS

- High vibration, thrust, temperature, and speed can occur during surge.
- Many instruments are too slow to detect surge quickly enough.
- Internal damage and loss in efficiency can result from surge.

QUESTIONS

1. Does the surge curve intersect the point where the slope of the compressor characteristic curve is positive, zero, or negative?

   ANSWER

2. Can compressor surge or stall occur when the operating point is to the right of the surge curve?

   ANSWER

3. What happens to compressor flow at the start of the surge cycle?

   ANSWER

4. Why is it difficult for a feedback controller to get a compressor out of surge once surge has started?

ANSWER

5. What happens to the amplitude, the period, and the shape of the surge oscillations as the plenum volume is increased?

ANSWER

6. What happens to the amplitude, the period, and the shape of the surge oscillations as the compressor speed is increased?

ANSWER

7. What are some of the consequences of surge?

    ANSWER

8. Why do many instruments detect surge too late?

    ANSWER

Effect of Operating Conditions

4-1 SURGE CURVE PLOTTING METHODS

The surge curve, which intersects the point of zero slope on the compressor characteristic curves, for a variable-speed single-stage compressor is a parabolic curve approximated by the following equations (Ref. 12):

$$\Delta P = P \cdot \{(M \cdot u \cdot (\pi \cdot D \cdot X \cdot Q)^2)/(g \cdot R \cdot Z \cdot T \cdot y)+1\}^y - P \quad (4\text{-}1)$$

$$y = (k \cdot e)/(k\text{-}1) \quad (4\text{-}2)$$

$$X = N/Q \quad (4\text{-}3)$$

where:

D = diameter of the impeller (ft)

e = polytropic efficiency (0.64 to 0.84)

g = gravitational constant (115920 ft/min^2)

k = specific heat ratio of gas (about 1.4 for air)

M = molecular weight of gas (about 29 for air)

N = impeller speed (rpm)

P = suction pressure (psia)

ΔP = pressure rise developed by compressor (psi)

Q = volumetric flow at suction conditions (acfm)

R = universal gas constant (R = 1545)

T = suction temperature (deg R)

u = pressure coefficient (typically 5.5)

X = speed-to-flow ratio (rpm/acfm)

Z = average compressibility factor (about 1 for air)

y = polytropic exponent ratio (2.2 to 3.0 for air)

The surge curve calculated from Equation 4-1 is plotted on the compressor map as pressure rise (ΔP) versus suction volumetric flow (Q). The above equations are not accurate for variable guide vanes, high compression ratios, multiple stages, or axial compressors. However the shift in the surge curve predicted as a function of the operating conditions, which appear as input variables in the equations, is qualitatively applicable to most compressors. These equations will be used to generate plots in Sections 4-1 through 4-4 to show the shift in the surge curve due to changes in suction pressure, suction temperature, and specific heat ratio. Also, the comparison of Equation 4-1 with the equation for a differential head flowmeter on the compressor suction shows how such a measurement can simplify surge control.

$$h = \{(C \cdot M \cdot P)/(R \cdot Z \cdot T)\} \cdot Q^2 \quad (4\text{-}4)$$

where:

C = constant that depends on the beta ratio and units

h = differential head signal (inches w.c.)

M = molecular weight of the gas

P = inlet pressure (psia)

Q = volumetric flow at inlet conditions (acfm)

R = universal gas constant (R = 1545)

T = inlet temperature (deg R)

Z = compressibility factor

If Equation 4-4 is solved for Q and substituted into Equation 4-1:

$$\Delta P = \{(u \cdot (\pi \cdot D \cdot X)^2 \cdot h)/(g \cdot C \cdot y) + P\}^y - P \quad (4\text{-}5)$$

Notice that both Equations 4-1 and 4-4 have volumetric flow squared on the right side. If Equation 4-4 is solved for volumetric flow squared and the result is substituted into Equation 4-1, the resulting Equation 4-5 has differential head instead of flow squared on the right side. If the surge curve is then plotted as compressor pressure rise (ΔP) versus flowmeter differential head (h), the plot becomes nearly a straight line, whereas before it was a parabolic curve for ΔP versus Q.

Figure 4-1 shows how the surge curve becomes linearized. The linearized curve allows a simple ratio station to be used to generate a set point that is parallel to surge curve. The ratio station gain is set to match the slope of the surge curve and the bias is set to offset the set point curve to the right of the surge curve. The input to the ratio station is usually ΔP and its output is the remote set point for a flow controller whose measurement is h (Section 6-4 will explain this in greater detail). The user must take the compressor map, which is a plot of ΔP versus Q, and calculate and plot the surge curve as ΔP versus h so that surge set point can be drawn and the ratio station gain and bias settings can be graphically determined. If the flowmeter is close to the compressor suction so that the flowmeter inlet temperature and pressure are about equal to the compressor suction temperature and pressure, the temperature and molecular weight terms and one pressure term cancel out. Thus the shift in the surge curve for a change in temperature, pressure, or molecular weight is less on a plot of ΔP versus h. In real compressor applications, these terms never completely cancel.

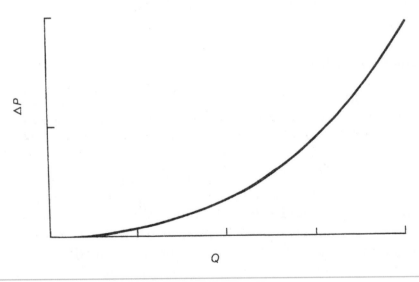

Figure 4-1a   Parabolic Surge Curve (ΔP versus Q)

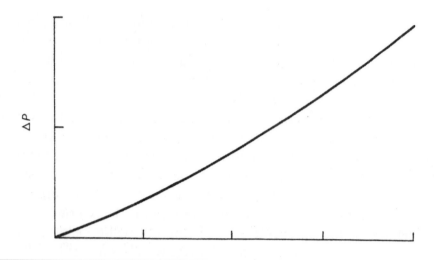

Figure 4-1b   Linear Surge Curve (ΔP versus h)

Key Concepts

For a variable-speed single-state centrifugal compressor:

- The surge curve is a parabolic curve on a plot of ΔP versus Q.
- The surge curve is a straight line on a plot of ΔP versus h.
- The changes in temperature and molecular weight cancel out.

4-2 SUCTION PRESSURE

Figure 4-2 shows that a *decrease* in suction pressure *lowers* the surge curve on a plot of pressure rise (ΔP) versus suction flow (Q). The operating point is now closer to the surge curve. This is caused by a downwards shift of the compressor characteristic curves. The surge curve is lowered by 20 percent for a 20 percent decrease in pressure at 80 percent flow. Equation 4-1 shows that the pressure rise is proportional to the suction pressure. A 20 percent change in suction pressure is much larger than typically experienced, especially if the suction pressure is atmospheric (remember that the suction pressure is in psia). A ΔP versus h plot of the surge curve would show that the decrease in suction pressure slightly raises the surge curve. Equation 4-5 shows that the pressure rise is proportional to the suction pressure to the $y-1$ power. Thus the magnitude of this shift in the surge curve depends

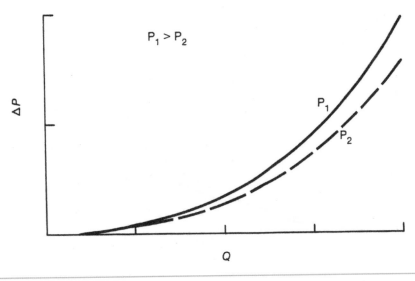

Figure 4-2   Effect of Suction Pressure

on the specific heat ratio and the polytropic efficiency. The value of y is always greater than 1, so that the shift is always upward for a suction pressure decrease.

Key Concepts

For a decrease in suction pressure of a variable-speed single-stage centrifugal compressor:

- The surge curve is lowered for a plot of ΔP versus Q.
- The surge curve is raised for a plot of ΔP versus h.

4-3 SUCTION TEMPERATURE

Figure 4-3 shows that an *increase* in suction temperature *lowers* the surge curve on a plot of pressure rise (ΔP) versus suction flow (Q). The operating point is now closer to the surge curve. This is caused by a shift downwards of the compressor characteristic curves. The surge curve is lowered by 21 percent for a 20 percent increase in suction temperature at 80 percent flow. Equation 4-1 shows that the pressure rise is inversely proportional to the suction temperature raised to the y power. Thus the magnitude of the shift also depends on the specific heat

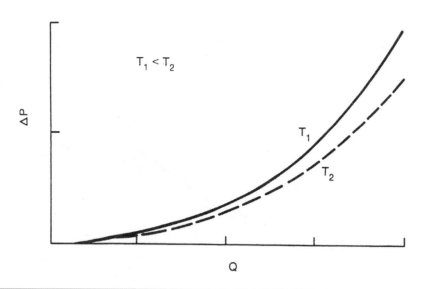

$T_1 < T_2$

ΔP

T_1

T_2

Q

Figure 4-3   Effect of Suction Temperature

ratio and the polytropic efficiency. A 20 percent increase in suction temperature is about the maximum change that can be experienced from winter to summer operation (remember that the suction temperature is in degrees R). A ΔP versus h plot of the surge curve would show that the change in suction temperature does not shift the surge curve. The temperature term has been cancelled out in Equation 4-5. In actual applications, the surge curve will shift due to changes in the specific heat ratio and inaccuracies in the equations.

Appendix C has P versus Q compressor maps that show the shift in the surge curve and surge set point of a variable guide vane multistage axial flow compressor for winter, spring, and summer operation. The surge set point is for a controller whose measurement is differential head of a suction flow measurement whose signal shifts with temperature per Figure C-5 in the Appendix. If the flowmeter signal shift had been equal to the surge curve shift so that the terms would cancel in Equation 4-5, the distance between the surge curve and the surge set point would have remained constant. Figures C-1 through C-3 show that the shift is in the same direction but not equal.

Key Concepts

For an increase in suction temperature of a variable-speed multistage centrifugal compressor:

- The surge curve is lowered for a plot of ΔP versus Q.
- The surge curve stays about the same for a plot of ΔP versus h.

4-4 MOLECULAR WEIGHT

Figure 4-4 shows that a *decrease* in molecular weight *lowers* the surge curve on a plot of pressure rise (P) versus suction flow (Q). The operating point is now closer to the surge curve. This is caused by a shift downwards of the compressor characteristic curves. The surge curve is lowered by 25 percent for a 20 percent decrease in molecular weight at 80 percent flow. A 20 percent change in molecular weight is larger than typically experienced unless different gases are blended or separated in the feed or recycle to the compressor. A plot of ΔP versus h of the surge curve would show that the change in molecular weight would not shift the curve. In actual applications, the surge curve will shift due to changes in the specific heat ratio and inaccuracies in the equations. Changes in specific heat ratio typically occur simultaneously with changes in molecular weight. The specific heat ratio increases as the molecular weight decreases for hydrocarbons, which will cause the surge curve to shift on a plot of ΔP versus h.

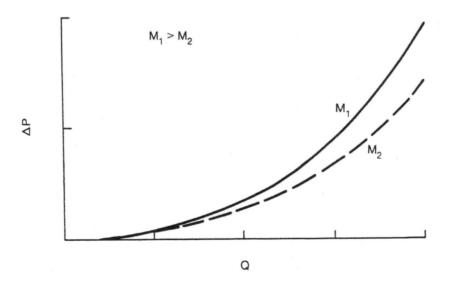

Figure 4-4   Effect of Molecular Weight

Key Concepts

For a decrease in molecular weight of a variable-speed single-stage centrifugal compressor:

- The surge curve is lowered for a plot of ΔP versus Q.
- The surge curve shifts for a plot of ΔP versus h for hydrocarbons.
- The specific heat ratio increases for hydrocarbons.

4-5 SPECIFIC HEAT RATIO

Figure 4-5 shows that a *decrease* in specific heat ratio *lowers* the surge curve on a plot of pressure rise (ΔP) versus suction flow (Q). The operating point is now closer to the surge curve. This is caused by the downward shift of the compressor characteristic curves. The surge curve is lowered by 9 percent for a 20 percent decrease in specific heat ratio at 80 percent flow. A 20 percent change in specific heat ratio is larger than typically experienced unless different gases are blended or separated in the feed or recycle to the compressor. A ΔP versus h plot of the surge curve would show that the surge curve is also lowered. The magnitude of the shift is the same for both types of surge curve plots even though the shift may look larger for the ΔP versus Q plot due to curvature. Whether the surge curve

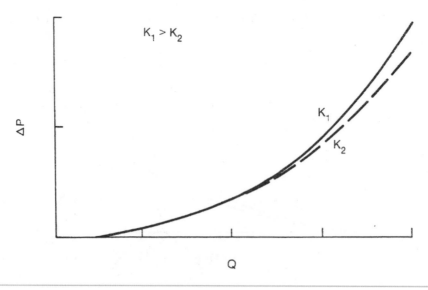

Figure 4-5   Effect of Specific Heat Ratio

shifts in the same or the opposite direction of the specific heat ratio depends on the particular values in Equation 4-1, because the two y terms that depend on the specific heat ratio have opposite effects. For an increase in y (decrease in specific heat ratio), the y exponent term acts to increase ΔP, while the y denominator term acts to decrease ΔP.

Key Concepts

For a change in specific heat ratio of a variable-speed single-stage centrifugal compressor:

- The surge curve is shifted for a plot of ΔP versus Q.
- The surge curve is shifted for a plot of ΔP versus h.
- The direction of the surge curve shift depends on the application.

4-6 COMPRESSION RATIO

Figure 4-6 is a plot of the compressor characteristic curves, the surge curve, and the choke curve for a variable-speed multistage compressor. Whereas the surge curve intersects the point of zero slope on the characteristic curve, the choke curve intersects the point of infinite slope (vertical line) on the characteristic curve. The flow cannot change at a given speed for operating points below the choke curve. This condition is also referred to as stonewall. Figure 4-6 shows that the addition of a compressor stage can cause a break point in the surge curve. The compression ratio (ratio of discharge pressure to suction pressure) of most compressors is limited to 3:1, while most applications require ratios of 3:1 to 15:1. Consequently, multistage compressors are quite common. As the number of stages and hence the number of break points in the surge curve increase, the curve develops a curvature opposite to that for the single-stage compressor.

Equation 4-1 should be modified so that the pressure rise (ΔP) varies with the square root instead of the square of the suction flow (Q) for a multistage compressor. The bending over of the surge curve means that the use of the flowmeter differential, which varies with the square of the flow, will now make the surge curve more nonlinear. Thus the remote set point for a suction flow controller should use the square of the discharge pressure, or the controller measurement should use a signal characterizer (see Section 6-3 for more details). The use of adjustable guide vanes can cause a similar bending over of the surge curve.

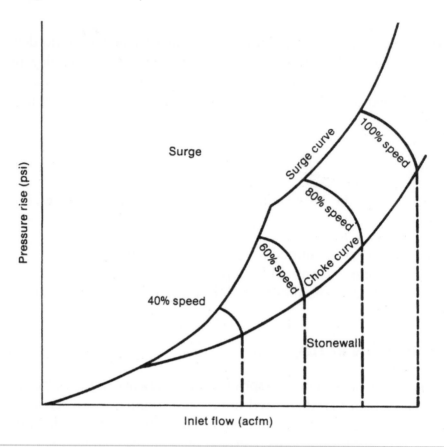

Figure 4-6   Effect of Compression Ratio

Key Concepts

- The surge curve bends over for a multistage compressor.
- The surge and choke curves can intersect for a multistage compressor.
- Surge prevention is difficult below the intersection point.

4-7 SPEED

Figure 4-7a for an axial compressor and Figure 4-7b for a centrifugal compressor show a family of speed characteristic curves. The slopes of these curves are much steeper for the axial compressor than for the centrifugal compressor. The change in flow due to a change in system resistance (caused, for example, by fouling or throttling) is less for an axial compressor. Also the process gain (percent flow change divided by percent speed change) is lower so that the speed regulation and variation errors correspond to a smaller error in flow (see Section 5-4 for more details). Thus an axial compressor is better for constant flow control and a centrifugal compressor is better for constant pressure control.

The steep negative slope for the axial compressor facilitates a more stable operation closer to the surge curve. The operating point will be more vigorously returned to the original point of the intersection of the load curve and the characteristic curve, and it will not wander into the surge region for small disturbances. However, notice that for constant flow, a speed increase due to the closing of a downstream valve will drive the operating point further into surge for an axial compressor. This action is graphically illustrated in Figures 4-7a and b by the

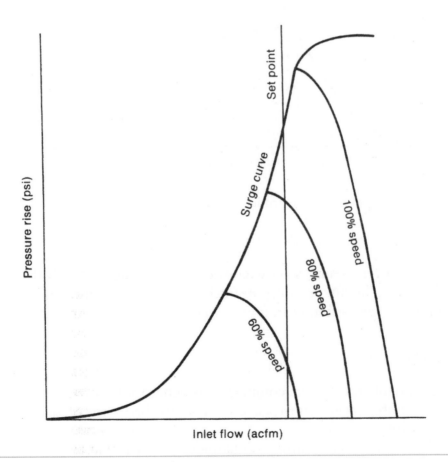

Figure 4-7a   Effect of Speed (Axial Compressor)

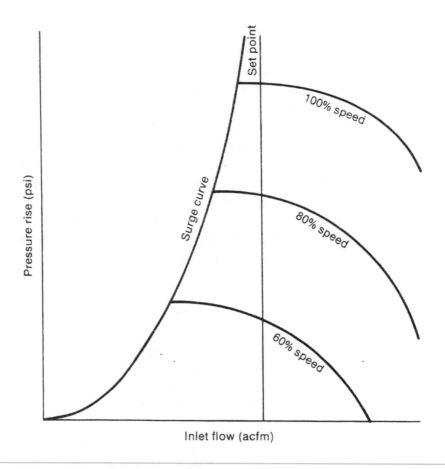

Figure 4-7b   Effect of Speed (Centrifugal Compressor)

vertical line that represents the flow control set point and the path the operating point will follow. This interaction, plus the more severe consequences of surge in an axial compressor, necessitates careful design of the surge and throughput controls.

Key Concepts

- A variable-speed axial compressor is better for flow control.
- A variable-speed centrifugal compressor is better for pressure control.
- Speed and surge control interaction is greater for an axial compressor.

4-8 VANE POSITION

Figure 4-8a for an axial compressor and Figure 4-8b for a centrifugal compressor show a family of vane characteristic curves. The slopes of these curves are steeper for the axial compressor than for the centrifugal compressor. Therefore, the conclusions from Section 4-7 on flow and pressure control performance and flow and surge control interaction for an axial compressor versus a centrifugal compressor also apply here.

Figures 4-8a and b also show that the surge curve bends over. Therefore, the conclusions from Section 4-6 on the need to either square the ΔP signal or use the signal characterizer for the h signal in order for the surge controller to linearize the surge curve also apply here.

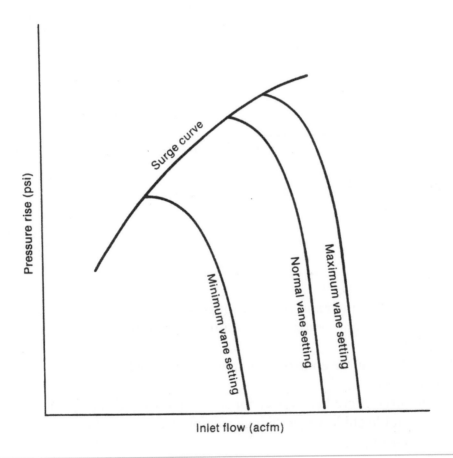

Figure 4-8a   Effect of Vane Position (Axial Compressor)

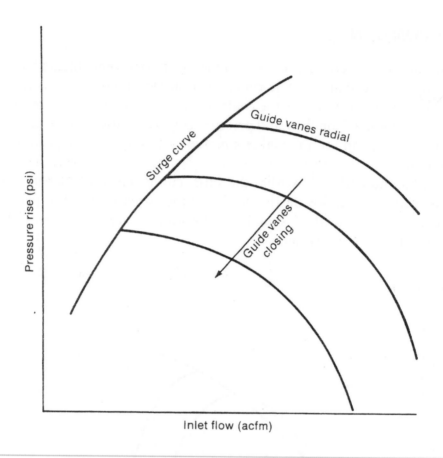

Figure 4-8b   Effect of Vane Position (Centrifugal Compressor)

Key Concepts

- A variable vane axial compressor is better for flow control.
- A variable vane centrifugal compressor is better for pressure control.
- Speed and surge control interaction is greater for an axial compressor.
- The surge curve bends over for a variable vane compressor.
- Adjustable guide vanes change suction pressure and gas prerotation.

QUESTIONS

1. Why are the percent changes in suction pressure less than the percent changes in suction temperature for an atmospheric suction?

   ANSWER

2. If the flowmeter for the surge controller were located on the discharge instead of suction, would the surge set point shift with temperature?

   ANSWER

3. Why is the surge curve more nonlinear on a plot of ΔP versus h than on a plot of ΔP versus Q for a compressor with multiple stages or adjustable guide vanes?

   ANSWER

4. How do adjustable guide vanes change the flow capacity of a compressor?

   ANSWER

Throughput Control

5-1 DISCHARGE THROTTLING

Figure 5-1 shows the use of discharge throttling to control the suction flow or the discharge pressure of the compressor. To decrease suction flow or increase discharge pressure, the discharge valve would be stroked further closed. This corresponds to the demand load curve raising, and the intersection of this curve and the compressor characteristic curve moving up from L1 to L2 in Figure 2-5 for a centrifugal compressor and in Figure 2-7 for an axial compressor. The operating point moves along a single characteristic curve. Thus the flow cannot be decreased or the pressure increased too much without the operating point approaching too close to the surge curve. Limits would have to be placed on either the controller output or the valve position to prevent the valve from closing too much and causing surge. Discharge throttling is also the least efficient of the throughput control methods because the energy loss is large due to the large pressure drop across the valve. Discharge throttling is rarely used due to this poor turndown and efficiency.

Key Concepts

- Discharge throttling has the least turndown capability.
- Discharge throttling is the least efficient throughput control method.

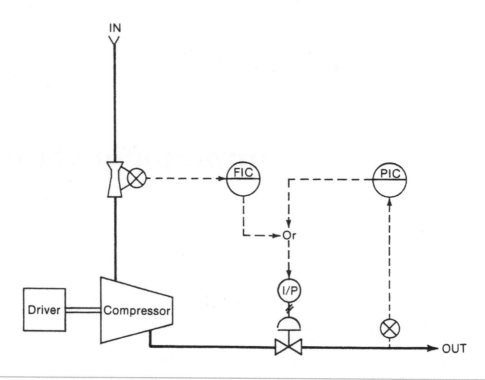

Figure 5-1   Discharge Throttling Schematic

5-2 SUCTION THROTTLING

Figure 5-2 shows the use of suction throttling to control the suction flow or the discharge pressure of the compressor. To decrease suction flow or decrease discharge pressure, the suction valve would be stroked further closed. This corresponds to the operating point moving to a lower characteristic curve on a plot of discharge pressure versus suction flow. The compressor map would resemble Figure 4-7a for axial compressors and Figure 4-7b for centrifugal compressors except that each characteristic curve corresponds to a different suction valve position instead of to speed, and the Y axis is discharge pressure instead of pressure rise. Because the operating point moves to lower characteristic curves, greater turndown of capacity is realized, since the positive slope of the surge curve ensures that the surge point will always be at a lower flow for a lower characteristic curve. The pressure drop across the suction valve is much smaller than that across the discharge valve; and the closing of the suction valve reduces the inlet gas density,

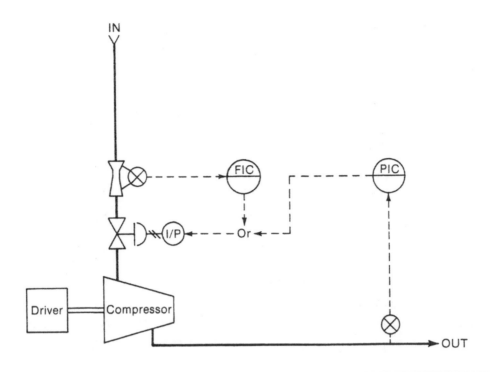

Figure 5-2   Suction Throttling Schematic

which reduces the horsepower requirement. Consequently, suction throttling is more efficient than discharge throttling.

Key Concepts

- Suction throttling has greater turndown than discharge throttling.
- Suction throttling is more efficient than discharge throttling.

5-3 GUIDE VANE POSITIONING

Figure 5-3 shows the use of guide vanes to control the suction flow or the discharge pressure of the compressor. To decrease suction flow or decrease discharge pressure, the guide vanes should be stroked further closed. This corresponds to the operating point moving to a lower characteristic curve in Figure 4-8a for axial compressors and Figure 4-8b for centrifugal compressors. A change in the guide

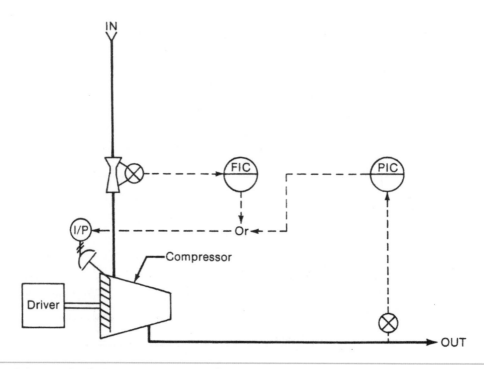

Figure 5-3   Guide Vane Positioning Schematic

vane position changes the suction pressure and the amount of prerotation of the gas. The vanes provide a maximum counterrotation of the gas at their maximum position and a maximum prerotation of the gas at their minimum position. The use of prerotation increases the efficiency at low loads. Guide vane positioning is more efficient than suction throttling and has the greatest turndown capability of all the throughput control methods.

Key Concepts

- Guide vane positioning has the greatest turndown capability.
- Guide vane positioning is more efficient than suction throttling.

5-4 SPEED CONTROL

Figure 5-4 shows the use of speed control to control the suction flow or the discharge pressure of the compressor. The speed control loop is the inner loop of a cascade control system. The set point of the speed controller is the output of either the flow or the pressure controller. The type of speed control depends on whether the driver is a turbine or a motor. The output of the speed controller can either throttle the steam or gas flow to a turbine, as depicted in Figure 5-4, or vary the electrical frequency of the power to the motor. To reduce the suction flow or the discharge pressure, the speed would be reduced by decreasing the turbine inlet flow or the motor power frequency. This corresponds to the operating point moving to a lower characteristic curve in Figure 4-7a for axial compressors and Figure 4-7b for centrifugal compressors. The power requirement is proportional to the cube of the speed if changes in efficiency are neglected. Therefore, the reduction in throughput by a reduction in speed is the most efficient method. Since the surge curve bends up for speed control and over for vane position control, the operating point is closer to the surge point at low flow, and the turndown capability of speed control is less than that for guide vane positioning. The speed control range cannot be extended so low that it includes critical speeds. The operating point must pass through critical speeds as fast as possible on startup to avoid damage due to high vibration.

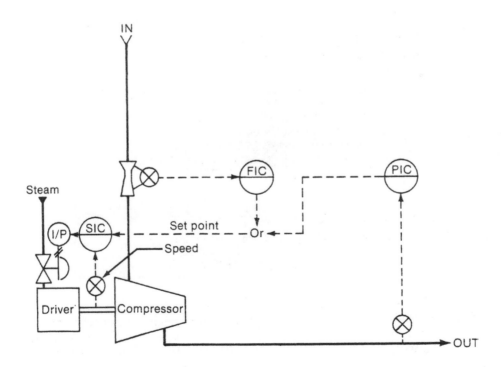

Figure 5-4   Speed Control Schematic

Figure 5-5 shows the operating range of compressor speed in rpm and of compressor inlet flow in acfm for different types of drivers. Whether a motor or a turbine driver is used depends also upon the availability and the cost of steam or gas versus power within the plant.

The sophistication of the instrument hardware used for driver speed control depends on the control loop performance required. Driver speed regulation and variation errors translate to a compressor flow or pressure control error. The compressor control error increases with process gain for a given speed control error. The process gain is the change in compressor flow or pressure divided by the speed change. The process gain can be estimated from the compressor map by drawing a horizontal line for a flow change at constant pressure or a vertical line for a pressure change at constant flow. This line starts at one speed characteristic curve and ends at another. The process gain is the length of the line in flow or pressure units divided by the difference in speed between the two characteristic curves. If the distance between the characteristic curves is larger than the normal operating

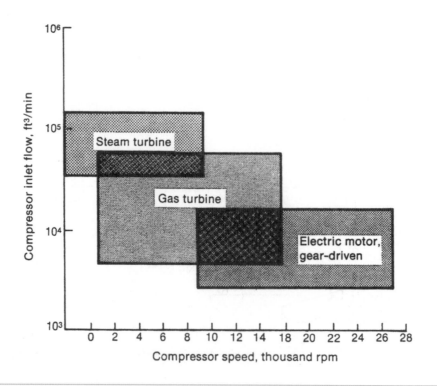

Figure 5-5   Compressor Driver Operating Ranges

range, an intermediate characteristic curve should be sketched. The flow process gain is large for centrifugal compressors and the pressure process gain is large for axial compressors, since the characteristic curves are flat for centrifugal compressors and steep for axial compressors. Thus centrifugal compressors are better for pressure control and axial compressors are better for flow control.

Table 5-1 Speed Governor Classification

Governor class	Regulation	Variation
A	10%	3/4%
B	6%	1/2%
C	4%	1/4%
D	1/2%	1/4%

Table 5-1 shows the speed regulation and variation errors for two speed governor classes. The speed regulation error is the difference between the speed at zero power output and the speed at rated power divided by the speed at rated power and multiplied by 100 percent. The speed variation error is the difference between the change in speed above set point and the change in speed below set point divided by the speed at set point and multiplied by 100 percent. The speed change is the difference between the speed with and without the speed governor in service. Speed variation includes dead band and sustained oscillations.

Electronic governors have been developed that offer the following advantages over mechanical and pneumatic governors (Ref. 30).

- Quicker speed response to load changes.
- Better environment and maintenance access by remote mounting.
- Greater speed turndown.
- Simpler interface with electronic throughput or surge controls.
- Simpler interface with computers.
- Reduced speed dead band by elimination of mechanical parts.
- Reduced maintenance from wear by elimination of mechanical parts.

The above list of advantages is formidable. However, the electronic equipment involved is relatively complex and new. While the electronic equipment is reliable if protected from the environment, it will require troubleshooting at times even if only to verify that it is functioning properly when control loop problems occur. The time and cost of training maintenance technicians should be considered.

Figure 5-6 shows the difference in system components between a mechanical-hydraulic system and an electronic-hydraulic system for speed control. The electronic system eliminates the many feedback links that require alignment, create dead band, and deteriorate from wear, corrosion, and vibration. The General Electric MDT-80® system shown also has redundant power supplies and speed sensors and 2-out-of-3 logic for an overspeed trip. The Tri-Sen Systems M-300® electronic governor provides speed regulation that is typically 0.1 percent of speed and has an optional digital T-111 tachometer that has an accuracy of 0.01 percent ± one rpm (see Section 7-3 for more details on speed measurement accuracy).

Throughput control errors decrease as the response time of the speed control loop decreases not only because the speed control loop can then correct for disturbances before they affect the throughput but also because the outer loop throughput controller need not be detuned if the inner loop is faster than the outer loop for a cascade system (Ref. 21). However, the interaction between throughput and surge control increases if the speed control is as fast as the surge control. This interaction problem can be solved by the addition of a decoupler between the throughput and the surge controller output (see Section 9-2 for more details).

The response time of the speed control loop depends upon the response time of sensor, controller, actuator, steam flow or gas flow or power hertz, and rotor. The time constants less than the largest time constant in the loop can be summed with the time delays and multiplied by 4 to estimate the loop period. The response time of the speed sensing and the electronic calculations is negligible and is only slowed down by filtering or by sampling. The response time of hydraulic positioning for steam or gas throttling is also negligible. The response time for pneumatic positioning is *not* negligible and increases with the size of the actuator. If a conventional diaphragm-actuated control valve is used to throttle the gas or steam flow, the valve may require 2 to 8 seconds to stroke full scale. The response time of the exit steam flow after a change in the inlet steam flow is fast and has been measured to have a time constant of only 0.1 second. The response time of power hertz is usually rate-limited to 3 or more seconds for a full-scale speed change to prevent motor torque overload. This response time can be reduced to 0.1 second or less by increasing the motor size and adding tachometer feedback. The response time of rotor speed to a change in driver torque can be estimated by the following equation (Ref. 31):

$$(I/K) \cdot (\Delta N/\Delta t) = (T_d - T_l) \quad \text{Momentum balance} \quad (5\text{-}1)$$

$$K = (c \cdot g)/2\pi \quad (5\text{-}2)$$

Mechanical-hydraulic multi-variable
control system

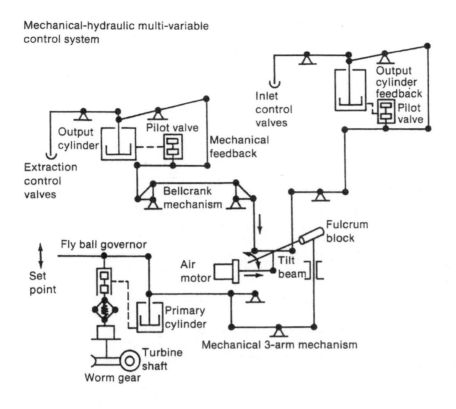

G.E. electro-hydraulic system

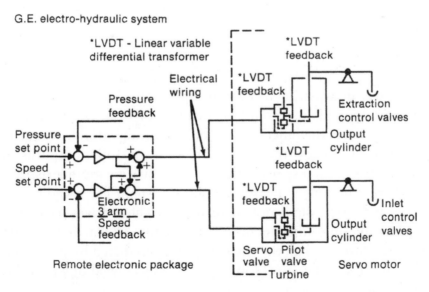

Figure 5-6　　Mechanical-Hydraulic and Electronic-Hydraulic Systems
(Courtesy Mechanical Drive Turbine Department, General Electric Company)

$$T_d = (5252 \cdot H_p / N) \cdot O_f \quad (5\text{-}3)$$
$$T_l = (F \cdot H) / (2\pi \cdot N \cdot e) \quad (5\text{-}4)$$

For $T_l = 0$:

$$\Delta t = (I/K) \cdot (\Delta N / T_d) \quad \text{Start-stop time} \quad (5\text{-}5)$$

where:

c = minutes-to-seconds conversion (60 sec/min)

e = efficiency

F = mass flow (lb m/min)

g = gravitational constant (32.2 ft/sec² lb ft/lb m)

H = head (ft · lb ft/16 m)

H_p = driver horsepower (hp)

I = inertia of driver and load rotor (lb m · sq ft)

K = units conversion factor (308 1/min · ft/sec · lb f/lb m · radians/rev)

N = speed (rpm)

O_f = overload factor

t = time (sec)

T_d = driver torque (ft · lb f)

T_l = load torque (ft · lb f)

The horsepower in Equation 5-3 can be approximated as a linear function of steam or gas flow (pounds per hour) for turbines or of power frequency (hertz) for motors. At low loads the function may be nonlinear. Variable frequency drives without tachometer feedback (the flow or pressure controller output is converted to a frequency rather than a speed set point) do not have an inner speed control loop, so that there is no longer a cascade control system. The speed response time is much slower and speed slip is not corrected for. These drives have an acceleration and deceleration time adjustment to limit the rate of change of the frequency. If these times are set *more* than the start-stop time predicted by Equation 5-5, the response time of the speed control loop is needlessly slowed down. If these times are set *less* than the start-stop time predicted by Equation 5-5, the motor will slow down or shut down due to torque overload. The start-stop time can be reduced by increasing the motor's horsepower rating or overload factor.

Key Concepts

- Speed control has the greatest efficiency.
- Speed control has less turndown than guide vane positioning.
- Throughput control errors increase with process gain.
- Electronic governors have many performance advantages.
- Throughput control errors increase with speed loop response time.
- The turbine response time is small if the actuator is hydraulic.
- The motor response time is small if tachometer feedback is used.

QUESTIONS

1. List the throughput control methods progressing from least efficient to most efficient.

   ANSWER

2. List the throughput control methods progressing from least turndown to most turndown.

   ANSWER

3. Why does process gain affect throughput control errors?

    ANSWER

4. Why does speed control loop response time affect throughput control errors?

    ANSWER

5. What are some things that can cause a speed loop to be slow?

    ANSWER

Surge Control

6-1 MINIMUM FLOW CONTROL

Since surge is caused by insufficient flow, why not use a simple flow controller with its set point set at some minimum value to recycle or vent flow in order to always keep the compressor flow above some minimum? This control strategy is cheap and easy to understand for maintenance and operation. However, Figure 6-1 shows that the minimum flow represented by the vertical line cannot prevent the operating point from approaching too close to the surge curve at high pressures and forces the operating point to be unnecessarily far away from the surge curve at low pressures. If the operating point crosses the surge set point at low pressures, the vent or recycle valve will open even though the operating point may not be close to surge. Also since the efficiency ellipses typically have their long axis parallel to the surge curve with the innermost ellipse having the maximum efficiency, the normal operating point is forced to the low efficiency extremities. Thus energy is wasted due to both unnecessary recycle or vent flow and low efficiency operation.

Key Concepts

- Flow control causes reduced surge protection at high pressures.
- Flow control causes needless venting and recycle at low pressures.
- Flow control causes operation at low efficiency.

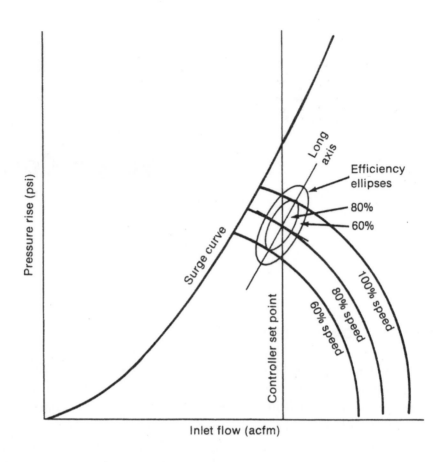

Figure 6-1   Minimum Flow Control Set Point

6-2 MAXIMUM PRESSURE CONTROL

Since surge typically occurs when a downstream block or throttle valve closes, which causes pressure to build up, why not use a pressure controller with its set point set at some maximum value to recycle or vent flow to relieve the buildup of pressure as a relief valve does? This control strategy is also cheap and easy to understand for maintenance and operation. However, Figure 6-2 shows that the maximum pressure set point represented by the horizontal line cannot prevent the operating point from crossing to the left of the surge curve at low pressures. If the operating point goes above the surge set point, the vent valve or recycle valve will open even though the operating point may be far to the right of the surge

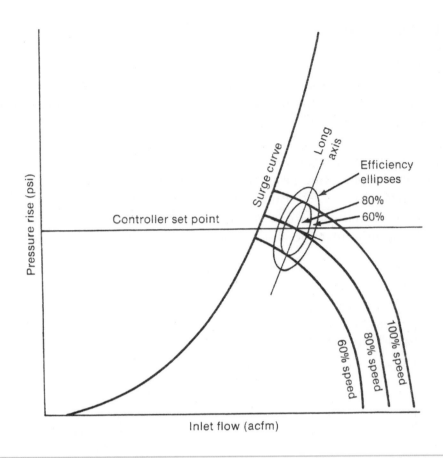

Figure 6-2   Maximum Pressure Control Set Point

curve. Also, since the efficiency ellipses typically have their long axis parallel to the surge curve with the innermost ellipse having the maximum efficiency, the normal operating point is forced to the low efficiency extremities. Thus energy is wasted due to both unnecessary recycle or vent flow and low efficiency operation.

Key Concepts

- Pressure control provides no surge protection at low pressures.
- Pressure control causes needless recycle and venting at low pressures.
- Pressure control causes operation at low efficiency.

6-3 RATIO CONTROL

Since the objective is to keep the operating point to the right of the surge curve, why not generate a set point for the surge controller that is parallel to the surge curve? This turns out to be the best control strategy. Not only is surge prevented for all possible positions of the operating point on the compressor map, but energy is saved by less unnecessary recycle or vent flow and higher operating efficiency. Since the long axis of the efficiency ellipses are parallel to the surge curve set point, the operating point moves along the long axis for load changes. Figure 6-3 shows that this set point is parallel to the surge curve and runs along the long axis of the ellipses for a surge curve that bends upward (single-stage compressor without adjustable guide vanes). In actuality the surge curve and set

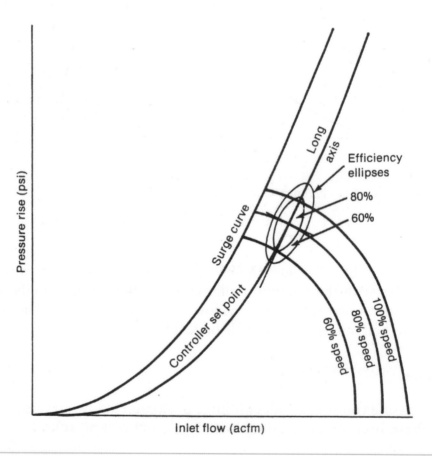

Figure 6-3   Ratio Control Set Point

point would be used on a plot of ΔP versus h for this application so that the surge curve approaches a straight line described by the following equation:

$$h = m \cdot \Delta P + b \quad (6\text{-}1)$$

where:

b = bias of surge set point

h = suction flowmeter differential head (inches w.c.)

m = slope of surge set point

ΔP = pressure rise developed by the compressor (psi)

Equation 6-1 shows the linear relationship between suction flowmeter differential head (h) and compressor pressure rise (ΔP) that was first documented by M.H. White of Foxboro (Ref. 29) and used extensively since then as the heart of proper feedback surge control. It doesn't matter which axis is used as the measurement signal for the surge controller. However, since it is somewhat more important for the operator to read the compressor flow than the compressor pressure rise, the X axis is assigned as the measurement signal and the Y axis is assigned as the set point signal of the surge controller. Figure 6-4 shows that the surge controller is then just a flow controller with a remote set point. The remote set point is the output of a ratio station whose gain is set to match the slope of the surge curve, whose bias is set to give the desired offset from the surge curve, and whose input is the pressure rise developed by the compressor. How large the offset needs to be depends on the speed of the instruments and the speed of the fastest disturbance that should not trip the open-loop backup system. (The open-loop backup is discussed in the next section.) It may be undesirable to activate the open-loop backup unless absolutely necessary in those applications where the compressor supplies multiple users, since the shutdown of one user would disturb the other users if the backup system overcorrects and opens the surge valves too much. If overcorrection is not a problem, then the addition of the backup system reduces the required offset of the feedback set point. If the instruments are fast, the feedback controller is tuned properly, and overcorrection is a problem, it is the downstream disturbance that determines the bias. If the downstream disturbance is the closing of a block or throttle valve, the installed characteristic of the valve should be plotted. The feedback surge controller and open-loop backup set points in flow are marked on the Y axis and are translated to stroke values on the X axis. The X axis is then graduated as to the downstream valve's stroking time so that the time interval between the feedback surge controller set point and

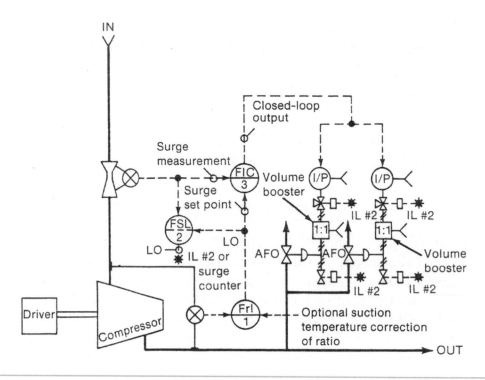

Figure 6-4   Surge Ratio Control Schematic

open-loop backup is determined. Appendix C shows an example of this technique for summer operation positions of the set points. The downstream block valve in this application was a butterfly valve that had a flat installed characteristic when the valve was greater than 75 percent open. The stroking time from 0 percent to 25 percent open was 120 seconds and the stroking time from 25 percent to 100 percent open was 20 seconds. This was considered to be so slow a valve that surge control should be easy. But notice that the actual time available between the feedback and backup set points is only *1.7 seconds* and between the backup set point and the surge curve is *1.1 seconds*. Whether the feedback and backup set points are far enough apart can be roughly checked by the following equation:

$$\Delta E = (\Delta V)/\{K_c \cdot (1 + 0.5 \cdot T_v / T_i)\} \quad (6\text{-}2)$$

where:

ΔE = distance between feedback and backup set points (percent)

K_c = controller gain (dimensionless)

T_i = integral time (sec/repeats)

T_v = stroking time of downstream block valve between set points (sec)

ΔV = change in surge valve position required (percent)

Equation 6-2 shows that the minimum distance between set points can be achieved only if the controller gain is maximized and the integral time is minimized. The maximum controller gain and minimum integral time are proportional to the response time of the instruments in the loop (see Section 12-1 for more details on tuning). A poorly designed surge control system may require the feedback set point to be 18 percent or more from set point, whereas a well-designed feedback plus backup surge control system may have its feedback set point less than 10 percent from the surge curve and the backup set point less than 5 percent from the surge curve. The energy savings increase as the distance of the set points from the surge curve decrease by reducing the unnecessary recycle and vent flow and by operating on the long axis of the efficiency ellipse, which is relatively close to the surge curve.

The flexibility of the signal characterizer and calculation blocks in a modern Distributed Control System (DCS) eliminates the need for the use of square root extractors to generate a set point curve parallel to the compressor surge curve. High compression ratios and multiple stages cause break points in the surge curve (see Section 4-6 for more details). The signal characterizer is used in place of the ratio station to match the curvature piecewise by the use of multiple line segments. Analog signal characterizers have been used in boiler control for years, but they require the adjustment of a gain and bias potentiometer for each line segment. Digital signal characterizers are more accurate and are more easily set up. To determine the gains and biases of the line segments, the compressor map should be rotated 90 degrees so that the X axis is pressure rise (input signal) and the Y axis is suction flow (calculated output). The line segments are then sketched in over the surge curve. Smaller line segments are used where the curvature is greatest, especially if it is near the intended operating range. The X and Y coordinates of the start and finish of each line are marked in percent of the suction flow and pressure rise transmitter ranges. The slope of that line segment is the difference in Y coordinates divided by the difference in X coordinates. If the gains are accumulative (the gain is not reset for each line segment), the incremental gain is the new slope minus the previous slope. The biases are the X coordinates at the start of each line segment. The overall output bias of the characterizer is adjusted to give the desired offset of the surge set point from the surge curve.

If the flow measurement is on the discharge instead of on the suction of the compressor, the discharge flow must be referenced back to suction temperature and pressure. The discharge and suction temperatures and pressures must be measured. The differential head signal is divided by discharge temperature, multiplied by discharge pressure, divided by suction pressure, and multiplied by suction temperature. This signal then passes through a signal characterizer and becomes the measurement for the surge controller. The discharge temperature measurement may not be fast enough to correct the flow signal for fast load changes because of heat transfer lags in the thermowell.

Key Concepts

- Ratio control provides the greatest efficiency and surge protection.
- The ratio station gain is set to match the surge curve slope.
- The ratio station bias is set to give the offset from the surge curve.
- The offset depends on the speed of the instruments and the disturbance.
- The surge controller is a flow controller with a remote set point.
- The remote set point is the output of the ratio station.
- A signal characterizer or calculation block can be used to match the shape of surge curves.
- A discharge flow measurement must be referenced to suction conditions.

6-4 OPEN-LOOP BACKUP

Once a compressor goes into surge, the feedback surge controller will probably not be able to bring the compressor out of surge since the surge oscillations are so fast that they are equivalent to uncontrollable noise to the surge controller. A fast opening but slow closing surge control valve will help prevent a second surge cycle. However, the additional lag introduced into the control loop oscillation by slowing down the valve in one direction necessitates the use of a smaller controller gain. Even though it has a well designed feedback surge controller, a compressor will be subjected to surges for the following reasons:

- The surge curve shape supplied by the compressor manufacturer is "expected" and not guaranteed.
- The surge curve shifts with wear and operating conditions.

- The compressor needs to be purposely surged to verify the surge curve.
- A downstream check valve may fail closed.
- A downstream block or throttle valve may fail closed.
- The surge controller may fail, be switched from auto to manual, or be switched from remote to local set point.

A loss prevention analysis would show that an open-loop backup is justified for all dynamic compressors. The open-loop backup detects too close an approach to surge or the actual start of surge and takes a preprogrammed corrective action. The operating point and flow derivative methods are the two major open-loop backup methods.

Figure 6-5a shows the operating point method. If the suction flow decreases more than a set percentage below the remote set point of the surge controller, a preprogrammed signal is generated by FY-4 to open the vent or recycle valve. Figure 6-5a shows an external deviation switch FSL-2 to activate a contact input of the FY-4 signal generator when the differential head signal falls more than a set percentage below the output of the ratio station FrI-1. This method detects when the operating point has crossed over the surge controller set point and is between this set point and the surge curve. It is better than the flow derivative method in that it will take corrective action before the surge curve is reached and is able to prevent the start of surge if properly designed. However, the backup protection is lost if the ratio station output goes to zero due to a pressure rise transmitter or ratio station failure.

Figure 6-5b shows the flow derivative method. If the suction flow drops rapidly, the output of the lead-lag station FY-6 goes from zero percent (zero derivative for steady flow) to some large output (large derivative for rapid flow drop). The lead-lag station gain is set for reverse action, its lag time constant is set to prevent noise from changing the lead-lag station output, and the lead time is set to give a large output change for a rapid flow drop. If the output is above a set limit, switch FSL-2 activates a contact input of the signal generator FY-4. This method detects the precipitous drop in flow that signifies the start of surge. This method takes corrective action after the start of surge and will usually prevent a second surge cycle if properly designed. It does not depend on having a pressure rise signal and therefore is useful for compressors with multiple stages and recycle flows where the pressure rise across a stage cannot be measured. If the flow measurement is too noisy, the lag setting required may be so large that it slows down the response of the lead-lag station too much to prevent a second surge cycle.

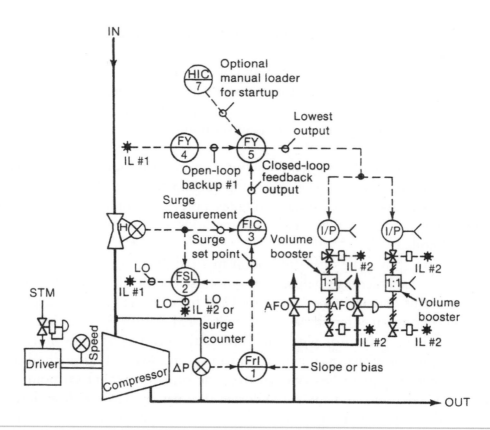

Figure 6-5a Open-Loop Backup Schematic—Operating Point Method (A smart digital positioner with a booster and properly set bypass on the positioner output is used instead of an I/P and booster for reasons described that support Figure 7-1b)

Regardless of what type of surge detection is used, the preprogrammed action should stroke the surge control valve open in less than a second and then allow a transition back to feedback surge control unless the compressor is to be immediately shut down. The transition is accomplished by dissipating the preprogrammed signal. The feedback surge controller output signal and the open-loop backup preprogrammed signal both are inputs to a signal selector, which selects the signal that demands the surge control valves to be the furthest open. Since the surge control valves are fail open, a low signal selector is used. The signal generator and surge controller outputs are then at 100 percent when the operating point is to the right of the surge curve (this corresponds to the measurement being above the remote set point of the feedback surge controller). The signal generator output

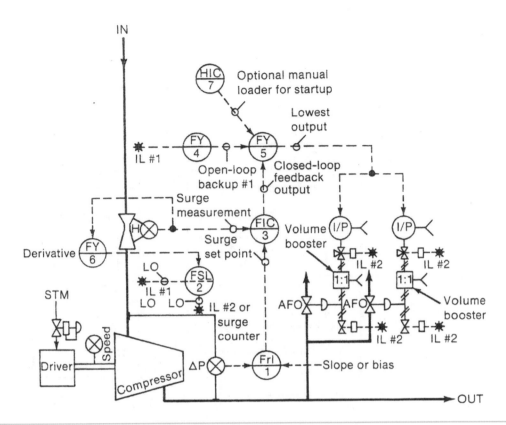

Figure 6-5b Open-Loop Backup Schematic—Flow Derivative Method (A smart digital positioner with a booster and properly set bypass on the positioner output is used instead of an I/P and booster for reasons described that support Figure 7-1b)

immediately goes to zero when activated and then slowly increases. Failure of the surge controller or the signal generator will cause the surge control valves to open. All of these functions can be done within a single microprocessor controller if it is fast enough.

Figure 6-6 shows a unique method of shifting the surge set point further to the right by adding a bias from block 7 to the fixed bias at block 2 when the measurement drops significantly below set point, which signifies the start of surge. The ratio station gain, which is the scaler input to block 1, remains fixed. The surge set point can be restored to its original value by pushing the reset pushbutton. The curve generator shown as block 5 corresponds to the signal generator previously mentioned. A summer is used instead of a signal selector. The

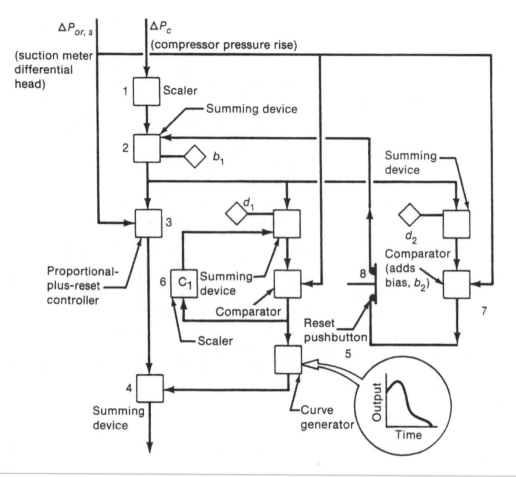

Figure 6-6 Automatic Surge Set Point Updating
(Courtesy Compressor Controls Corporation)

reversing of the signal for the reverse-acting surge control valves is not shown. Failure of the signal from the summer block 4 will leave the surge control valves in the closed position. The feedback surge controller (block 3) is a suction flow controller whose remote set point is the output of the ratio station (block 2). This system protects the compressor even though the surge curve may shift due to changes in operating conditions or wear.

The open-loop control response produces a rapid opening of the antisurge valve. The magnitude of this opening is adjustable through parameter C_1. Large values of C_1 will always protect the compressor in a properly designed system, even for very large disturbances. However, large values of C_1 tend to produce significant process disturbances, even process shutdowns. An additional parameter C_2 is used to reduce this problem. C_2 is the time delay between subsequent applications of C_1. That is, when the operating reaches the open-loop set point on the compressor map, C_1 is added to the output. The controller then waits through the C_2 time period. If the operating point is still on or to the left of the recycle trip line, C_1 is again applied to the output. This will continue until the operating point crosses the recycle trip line. The result of this combined action of C_1 and C_2 is that the compressor can be protected from relatively minor disturbances with a single application of C_1. The process disturbance is minimized. Large disturbances can be handled with cumulative applications of C_1. Typical values of C_1 and C_2 for a process sensitive to upsets would be C_1 from 15 to 25 percent and C_2 from 0.3 to 0.8 seconds. Naum Staroselsky received U.S. Patent 856,302 for this method and apparatus of combining closed-loop and open-loop control responses.

Key Concepts

- A feedback surge controller may not be able to stop a surge after it starts.
- A fast opening but slow closing surge valve may prevent a second surge cycle.
- Surges will occur even for a well-designed feedback surge controller.
- The operating point method tries to prevent the start of surge.
- The operating point method requires a reliable pressure measurement.
- The flow derivative method can prevent a second surge cycle.
- The flow derivative method requires a low-noise flow measurement.
- The surge set point can be updated on-line for a changing surge curve.

6-5 MANUAL AND PROCESS OVERRIDE CONTROL

The surge control valves may need to be throttled for a variety of reasons other than surge control. Regardless of the type of override controller, its output signal should go through a low-signal selector so that the surge controller and the open-loop backup can still open the surge valves if the operating point approaches the surge curve. The output signal demanding the largest opening of the surge control valve should always be selected. Sometimes a surge valve manual loader is installed at a local field panel beside the compressor and next to the displays of compressor lube oil pressure and temperature. This manual loader is sometimes used to position the control valves manually during startup. However, a properly designed surge control system should be able to automatically throttle the surge control valves as necessary during startup and provide better protection than a manual loader can. An operator cannot react as fast as the open-loop backup system. The manual loader output should go through the signal selector so that the operator can have manual control over the surge valve without disabling the feedback surge controller or open-loop backup. The operator can open the valve more but not less than what is demanded by the surge control system. If the operator forgets to increase the manual loader output to 100 percent (air fail open valve), the valve will stay open and considerable energy will be wasted.

Figure 6-7 shows the addition of a manual loader and a pressure override controller to the surge control system. The high-pressure override controller protects downstream equipment from high-pressure excursions. The pressure override controller should have external feedback. Its proportional-plus-reset control algorithm should be designed so that it does not demand the surge control valve to open until the discharge pressure rises above the override set point. The override set point should be drawn as a horizontal line on the compressor map and its position should be checked relative to the surge control set point and the operating point. If the override set point is within a few psi of the operating point, small pressure upsets will open the surge valves and waste energy.

Figure 6-8 shows a mass balance override control system to maximize reactor on-stream time. This override control system prevents the shutdown of one reactor from upsetting the mass balance in the header. The reactors have interlocks that shut them down if the gas feed flow is too high or too low to prevent flammable mixtures. Frequently these reactors exhibit a domino effect; the shutdown of one reactor will drop the header pressure enough to cause another reactor shutdown. If the reactors are exothermic and are a major steam supplier to the rest of the plant, multiple reactor shutdowns can cause a plant-wide shutdown. In this override control system, the decrease in flow to one reactor is immediately seen as a decrease in the total flow measurement (the compressor suction flow measurement will respond within milliseconds if a fast transmitter is used), while the total flow controller set point is still at its last value before the reactor shutdown. This is due to the lag imposed by the lead-lag station (the lead is turned off). The input to the lead-lag station is the sum of the reactor feed flows plus a small bias. In steady-state operation the set point is slightly above the measurement due to the bias so that the surge valves remain closed to save energy. When a reactor shuts down, the measurement

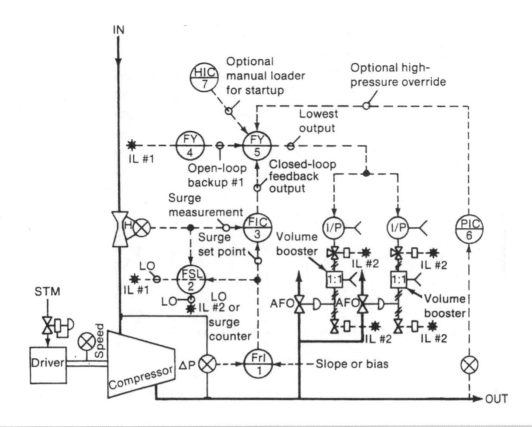

Figure 6-7 High-Pressure Override Schematic (A smart digital positioner with a booster and properly set bypass on the positioner output is used instead of an I/P and booster for reasons described that support Figure 7-1b)

drops below set point, and the override controller opens the surge control valves enough to bring the measurement back to the total flow before shutdown in order to maintain the mass balance in the header. Since the surge valves were designed to stroke in less than a second for surge control, the override control system is fast enough to maintain the mass balance. The total flow set point will gradually change to the new total flow for one less reactor, but the transition will be smooth and slow. Thus the surge control valves are slowly closed to save energy. This control system also protects other reactors against an operator lowering the set point of one reactor too quickly. It protects against the operator raising the set point of one reactor too rapidly only if the surge control valves are already open.

An "anticipator" control algorithm similar in strategy to the mass balance override control system was independently developed. The "anticipator" algorithm

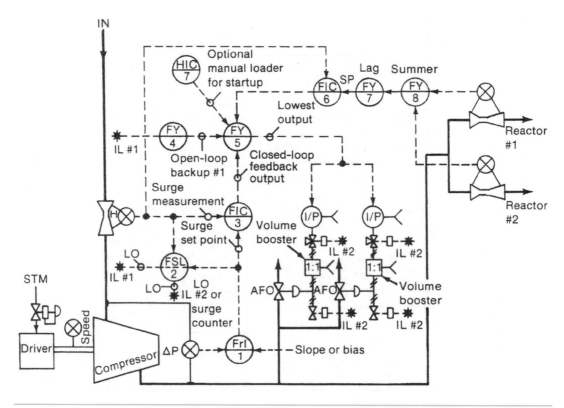

Figure 6-8   Mass Balance Override Schematic (A smart digital positioner with a booster and properly set bypass on the positioner output is used instead of an I/P and booster for reasons described that support Figure 7-1b)

uses a velocity limiter instead of a lag; it uses the surge measurement instead of the sum of user flows as the input to the velocity limiter; and it uses a high-signal selector instead of a separate flow controller. This signal selector chooses the largest output from the ratio station and velocity limiter as the set point for the surge controller. The algorithm is called an "anticipator" because it opens the surge valve before the operating point reaches the set point computed by the ratio station (Ref. 3).

If a condition develops that is known to cause a complete stopping of all forward flow from the compressor, an interlock can be used to open the surge valves immediately. For example, if all the block valves on a header distribution system close, block valve limit switch contacts wired in parallel would open and de-energize the solenoid valves to open the surge valves. This is shown as IL#2 in the control system schematics in this section. The use of this type of interlock allows the surge control system to be designed for a smaller disturbance, which allows set points closer to the surge curve for more efficient operation. If a solenoid valve with a large enough flow coefficient and the proper operating pressure range and materials of construction cannot be found, a ball valve actuated by a smaller solenoid valve can be used instead.

Key Concepts

- The override output should go through a low-signal selector.
- The manual override is not needed but is sometimes used.
- The high-pressure override protects downstream equipment.
- The mass balance override increases reactor on-stream time.

QUESTIONS

1. What are the disadvantages of minimum flow surge control?

   ANSWER

2. What are the disadvantages of maximum pressure surge control?

   ANSWER

3. Why is a square root extractor required if the surge curve on a plot of ΔP versus Q is linear?

ANSWER

4. What determines how close the surge controller set point can be to the surge curve?

ANSWER

5. Why is a low-signal selector used for override control?

ANSWER

6. What happens when one of the inputs to the signal selector fails downscale?

   ANSWER

STUDENT SUMMARY NOTES AND QUESTIONS FOR INSTRUCTOR

Instrument Requirements

7-1 CONTROLLER

The surge controller must have an *anti-reset windup* option. The surge control valves are closed and the measurement is above set point during normal operation. The surge controller is direct-acting since the surge valves are reverse-acting (air fail open) and the process is direct-acting (suction flow increases if the vent or recycle flow increases). The surge controller output is therefore full-scale. Since the flow is above set point with the surge control valves closed, the reset action of the controller integrates the positive error and saturates the output. When a user shuts down, the surge controller output remains saturated above 100 percent until the measurement is below set point (the operating point to the left of the set point) long enough for reset action to integrate the negative error and decrease the output to 100 percent. This time is so long that the operating point reaches the backup set point for even the slowest conceivable disturbance. The anti-reset windup option limits the contribution of reset to the controller output so that it plus the contribution of proportional action is equal to about 105 percent (the extra 5 percent signal is to ensure that the surge valves are tightly shut for normal operation). As soon as the measurement starts to decrease, the contribution of the proportional action to the controller decreases, the controller output decreases below 105 percent, and the surge valves start to open. If the approach to set point is slow, the increasing reset contribution to the output will cancel out the decreasing proportional contribution so that the output stays at 105 percent. Thus proportional action will open the surge valves before the measurement reaches set point for rapid disturbances. Some digital controllers have their algorithm deliberately designed to prevent the output from decreasing until the measurement is within 1/2 proportional band of the set point for even rapid disturbances. While this design is desirable for override control, it is undesirable for surge control (Ref. 21). If a reversing relay is used in the I/P transducer or valve positioner for the reverse-acting control valves, the surge controller action is reverse and the output is below 0 percent during normal operation (output is 0 mA dc for a 4 to 20 mA dc output signal). While windup is less severe in this direction, it is still a problem and the anti-reset windup option should still be used.

The controller should be electronic. If a digital controller is used, its sample time should be 0.3 second or less. If a flow derivative backup control system is implemented in the digital controller, the backup system may miss the precipitous drop in flow at the start of surge unless the sample time is 0.05 second or less. The digital controller will alias surge oscillations if the sample time is greater than 1/2 the oscillation period. The digital controller sample time and its associated filter add to the control loop a dead time equivalent to about the sample time. Since the total loop dead time determines the speed of the surge control system, the use of a slow digital controller must be compensated by the use of a faster transmitter and/ or control valve to prevent deterioration in the speed of the control system. If a fast transmitter is already being used, the control valve accessories or their piping must be optimized to reduce the stroking time for throttling.

The digital controller facilitates great flexibility in compensating for nonlinear surge curves, changes in operating conditions, and advanced control strategies. Look-up-tables, polynomial functions, and successive line segments can be used to generate a nonlinear set point that parallels almost any type of surge curve. Suction pressure can be measured for compressors in series and used to raise or lower the set point so that the distance between the set point and the surge curve remains constant. Almost any calculation made by the compressor manufacturer that is not iterative can be implemented. Iterative calculations are not allowed because recycling through an algorithm may require more time than the sample time. Function steps can be used to jump forward but not backwards in the program.

All of the control strategies described in this text can be implemented in a single powerful microprocessor controller. The elimination of multiple computing modules and their interwiring reduces installation and maintenance costs. Also, the control strategy can be easily revised during startup. The controller gain can be characterized so that the gain increases with deviation to the left of the set point on the compressor map if the operating point approaches too close to the surge curve. Controller manufacturers frequently refer to this feature as adaptive gain.

Key Concepts

- The surge controller must have anti-reset windup.
- The surge controller must be electronic.
- Digital controller sample time should be 0.3 second or less.

7-2 CONTROL VALVES

The most common mistake made during the design of a surge control valve installation is to ignore the need to throttle quickly in either direction. Frequently the valve accessories, such as solenoid valves and quick-exhaust valves, are installed to provide a fast full-scale stroke of 1 second or less but require 5 seconds or more to throttle to intermediate positions. The throttling speed depends on how much air can be moved into or out of the actuator for small changes in the surge controller output. Small orifice solenoid valves and small diameter tubing will restrict the air flow. Quick-exhaust valves will dump a large quantity of air for a large decrease in signal but act as restricters to air flow for a small decrease or for any size increase in signal. Since the surge control valves are air fail open, the quick-exhaust valve causes the surge control valves to open quickly but to close slowly. This stroking speed characteristic tries to duplicate the function of the open-loop backup signal generator. However, controller gain (K_c) and integral time (T_i) in Equation 6-2 increase as the throttle response time increases. Thus the quick-exhaust valve necessitates operation further from the surge curve and deteriorates controllability of the header mass balance, which is important for preventing the shutdown of one user from upsetting another user. It is best to design the surge control valve to respond quickly and precisely for small signal changes to ensure good feedback control; and it is best to design the signal generator to provide the desired stroking speed characteristic for preprogrammed open-loop action. The total of the control valve prestroke dead time and stroking time for throttling full scale should be less than 1 second.

For all but the highest speed and force requirements, pneumatic actuators instead of hydraulic, electro-hydraulic, motorized, or electromechanical actuators are used because of the following advantages:

- Lower Cost
- Fail Safe Action
- Safe for Hazardous Areas
- Simplicity
- Reliability
- Maintainability
- Standardization
- Minimal Parts Inventory
- Lower Energy Use

Figure 7-1a shows the use of a *volume booster* to increase the throttling speed of a control valve with a pneumatic actuator (a volume booster is a pneumatic relay with a large flow capacity). The booster mounting must be carefully designed to minimize the size and length of all air piping. The air supply regulator for the booster must have a flow capacity greater than that of the booster. The actuator air connection size should be enlarged. The spring side of a spring return actuator should have several vent holes to relieve the buildup of air pressure during the compression of the spring. If solenoid valves are required for interlocks, the exhaust solenoid valve should be 2-way with a large flow capacity close-nippled into a pipe tee at the actuator. A 3-way solenoid should block and vent the signal to the booster as the 2-way solenoid exhausts the actuator in order to prevent the booster from trying to fill the actuator. Boosters may not be required for control valves 4 inches or smaller that have small actuators. The booster dead band (smallest change of input signal that will change the output signal) may be 0.25 psi or larger even though the outlet sensitivity is 1 inch w.c. or better. A larger pneumatic signal range (i.e., 6 to 30 psi instead of 3 to 15 psi) should be used so that the dead band in percent of stroke is less. The high outlet sensitivity can cause butterfly disc flutter due to turbulence when the disc is near the wide-open position. If the butterfly shaft position can be manually moved near the wide-open position, the actuator spring rate must be significantly increased or a valve positioner must be added to prevent the disc flutter. The booster's high outlet sensitivity can exhaust the actuator for any slight compression of the actuator air loading from a slight upward movement of the shaft and thus allow the establishment of a new disc opening. The booster should have high inlet sensitivity and low outlet sensitivity for fast throttling of a control valve without a positioner.

Figure 7-1b shows the use of a volume booster to increase the throttling speed of a control valve with a *valve positioner* (a high-gain proportional-only controller whose measurement is valve position and whose set point is the I/P output).

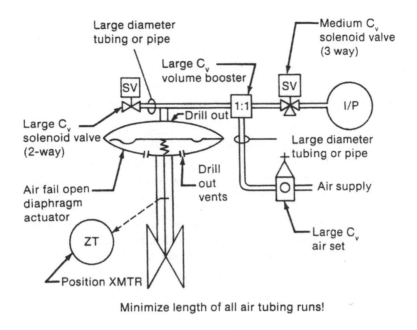

Figure 7-1a   Surge Control Valve Accessories for Fast Throttling (without a Positioner)

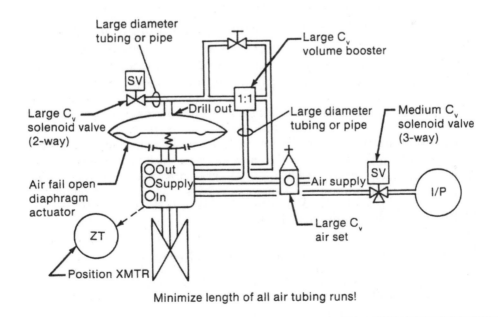

Figure 7-1b   Surge Control Valve Accessories for Fast Throttling (with a Positioner)

The positioner has a small flow capacity (small C_v) so that the throttling speed is slow even though the positioner will overdrive the output through its high gain. A valve positioner does not need reset action because the high gain reduces the offset to less than 1 percent. A booster can be added on the output of the positioner to speed up the valve, but the combination of a booster and a positioner in series is unstable unless a bypass with an adjustable restriction is added around the booster (Ref. 20). If the restriction is closed, the valve will go into a fast limit cycle whose amplitude is 20 percent or more and whose frequency is 1 hertz or more. The limit cycle occurs because the positioner is able to change the pressure in the small booster inlet volume faster than the booster can change the pressure in the large actuator volume. The addition of a booster in parallel in an application increased the volume seen by the positioner and allowed the restriction to be further closed. The improvement in response time through the use of boosters increases as the actuator volume increases and the restriction is closed. New valve positioners that have an adjustable dynamic gain may facilitate the closing of the bypass restriction. Control valves that are as large as 18 inches have been made to throttle to any position in less than 0.50 second by properly sized and installed boosters.

A pneumatic positioner will necessitates the detuning of an electronic analog surge controller because the positioner creates a cascade loop where the outer (electronic flow) loop is faster than the inner (pneumatic positioner) loop (Ref. 21). A smart digital positioner can be tuned to be faster than a Distributed Control System (DCS) loop. The faster digital positioner execution, smart tuning options, the use of predominantly reset action in the DCS flow loop, and the dynamic reset limit option with position readback for the PI controller makes the cascade loop concern a non-issue. A positioner is required when any one of the following conditions exist:

- Piston actuator
- Grafoil® packing
- Butterfly with disc flutter
- Butterfly with an eccentric disc and without torque compensation
- Booster with a large dead band

A smart digital positioner can do the following (Ref 30):

- Improve resolution (reduce stick-slip)
- Deal with diaphragm bench settings
- Deal with piston crossover pressure settings

- Increase actuation stiffness (prevent flutter)
- Increase or slow down valve speed of response
- Provide valve position monitoring and historization
- Provide valve alerts and diagnostic monitoring and historization

The Boosters in Figures 7-1a and 7-1b are used to increase the supply and exhaust flow to a pneumatic actuator to make it faster. The booster has a much higher flow capacity than an I/P or positioner to achieve and maintain the pressure in the actuator and hence the booster outlet at the pressure set based on the pressure signal at the booster inlet. The booster offers no stiffness beyond what is in the actuator. In several instances, high flow rate boosters on diaphragm actuators exhibited positive feedback (negative stiffness). This was observed in both the shop and field where a person grabbing the shaft of a 18 inch butterfly valve could manually move the disc when a booster was put on the output of an I/P and piped directly to an actuator. When the booster was removed (I/P output went to actuator) or was replaced with a positioner, the valve stem could not be moved. The high outlet port sensitivity of the booster caused the booster to relieve or supply pressure from an ever so slight compression or expansion of the diaphragm volume to provide a force in the same direction of the force applied to the stem (Ref 30).

Boosters have poor inlet (signal) port sensitivity. The booster will not respond to small changes. For slowly changing signals, a deadtime is added that is proportional to the booster deadband or resolution limit divided by the rate of change of the input signal. Since most controller output signals are making small changes from scan to scan, we have the ironic situation where a booster that is added to speed up the response of a valve actually prevents it from moving. The large step changes (e.g. > 5%) that are normally used to test valve response do not reveal the problem. The lack of understanding the implications of high outlet port sensitivity, poor inlet port sensitivity, bench settings, and fluid-forces has lead to the misguided rule that boosters instead of positioners should be used on fast loops (Ref 30).

A booster should not be used in lieu of a positioner but in conjunction with a positioner to maintain actuation stiffness, sensitivity, and a consistent input signal range for stroking the valve. The booster is mounted on each output port of the positioner. For double acting pistons, two boosters are required. The booster must have a bypass as noted in Figure 7-1b to prevent instability. Since the positioner is designed to be looking into a relatively large volume of an actuator, the extremely small volume of the booster inlet port will cause a rapid limit cycle. The bypass must be opened until the high frequency oscillations (e.g. 1-2 cps) stop. The bypass may be opened slightly further to provide a stability margin. Since the booster bypass slows down the speed of actuator, it is desirable to open the bypass just enough to provide stability. If the tuning of the digital positioner is changed, the booster bypass may need to be adjusted accordingly. More aggressive tuning settings will require a larger bypass flow (Ref 30).

The combination of a positioner and booster can dramatically reduce the pre-stroke deadtime and stroking time of large actuators. The air pipe or tubing size must be increased and in some cases the diaphragm actuator casing connection enlarged. The flow capacity of accessories, such as solenoid valves and air filter regulators, must be accordingly increased. Otherwise, the true flow capability of the booster is restricted (Ref 30).

When piston actuators are used, individual boosters with a restricted bypass are installed on each positioner output. Eccentric disc control valves have a non-linear torque requirement where the torque requirement plummets to a minimum immediately after the disc breaks away from the seat (Ref. 21). The result is delayed breakaway from the closed position and then overshoot. If this overshoot cannot be tolerated because it upsets the header mass balance, a positioner must be added to catch and recover from the overshoot. The valve recovery becomes faster as the booster capacity increases and as the bypass restriction is closed. Posi-Seal has developed a mechanical linkage to compensate for the torque nonlinearity.

An electronic valve position transmitter or a smart positioner with position readback is recommended because the valve speed deteriorates with age, and the stroking time is too fast to be quantitatively measured with a stop watch. The position transmitter output should be compared with the surge controller output on a high-speed recorder or data logger. The plot or log should be checked for the ability of the valve to respond precisely and quickly to small changes in controller output. Two surge control valves in parallel are recommended for additional reliability because the control valve with its accessories is the weak link in the system, and the valve cannot be maintained or tested on-line if it is the only one.

It is absolutely essential that the throttling speed and precision requirement be emphasized on the control valve specification, that the installation details of accessories be reviewed prior to assembly, and that the valve be throttle-tested and witnessed by the customer before shipment. An eccentric disc valve should be tested in a pipeline initially pressurized to the level of the compressor discharge because the upstream pressure acts to increase the sealing force and hence the breakaway torque and overshoot problem. The test should use a small ramping change in signal rather than a large step change in signal. The valve will respond much faster to a large step change, especially if the booster dead band is large. The control valve specification should contain a statement similar to the following: "The control valve should throttle to the correct position within 1 percent and in less than 1 second after a step change of 5 percent or more in I/P signal and throttle to track with less than a 1-second delay after a ramp change of 100%/sec or less in I/P signal."

A linear flow characteristic is preferable to an equal-percentage or quick-opening flow characteristic because the *inherent* characteristic becomes the *installed* characteristic due to critical flow. In typical flow control applications the variable pressure drop causes an equal-percentage characteristic to distort towards a linear characteristic and a linear characteristic to distort towards a quick-opening characteristic. A linear installed characteristic keeps the control loop gain linear so that the controller gain requirement does not change with control valve position. Otherwise the controller gain must be decreased to match the maximum valve gain, and the control action will be sluggish at other positions. Linear globe valves are recommended except when the line size is too large. Globe valves do not have the breakaway torque problem or the hysteresis problem that rotary valves have, and they are therefore able to throttle more precisely, especially near the closed position.

Each surge control valve should be sized large enough to pass the entire forward flow of the compressor with only 70 percent of the discharge pressure. The full discharge pressure is not available because it falls during the reverse flow cycle of surge. Each surge control valve should be sized small enough so that throttling near the closed position is not required. Valve positions of less than 10 percent should be avoided for eccentric disc valves.

A digital control valve, which contains individual solenoid actuated ports that increase in capacity per the binary code (2,4,8,16 etc.), has been used for surge control. The stroke precision can be estimated by the following equation (Ref. 21):

$$E = 100\%/2^n \quad (7\text{-}2)$$

where:

E = error in stroke

n = number of ports

A digital valve with 10 ports has a precision error of less than 0.1 percent. The stroking time is 50 to 100 milliseconds, the characteristic is linear, and the overshoot is zero. The life of the individual solenoids is prolonged by reducing the power input periodically (the duration of the reduction in power is not long enough to actuate the solenoid), which reduces their operating temperature. The life of the smallest solenoid and the port is prolonged by the use of a filter on the valve signal to prevent dither. However, there are some maintenance considerations. The sticking or failure of an individual actuator cannot be determined by external inspection, and the smallest port is susceptible to plugging with dirty gases.

Key Concepts

- Surge valves should throttle quickly and precisely.
- Boosters increase the stroking speed.
- Positioners reduce hysteresis, dead band, overshoot, and disc flutter.
- Boosters on the output of positioners need an adjustable bypass.
- Surge valves should have electronic position transmitters.
- Redundant parallel surge valves should be used.
- Surge valves should be throttle-tested and witnessed before shipment.
- The surge valve flow characteristic should be linear.
- Surge valves should pass 100 percent of the compressor flow at 70 percent pressure.
- Digital control valves are extremely fast and precise.

7-3 SPEED MEASUREMENT

The sensors for speed measurement are usually proximitor probes or magnetic pickups. A proximitor probe consists of a flat coil of wire coated with epoxy fiberglass on a ceramic tip. The proximitor probe is excited by an RF source. When a conductive surface comes close to the probe tip, eddy currents are set up on the surface and power is absorbed. How much power is absorbed depends on the distance of the surface from the tip. When a wheel with teeth is rotated in front of the probe, the probe output voltage drops each time a tooth passes by. A magnetic pickup consists of a permanent magnet surrounded by a coil of wire. When a magnetic surface passes near the tip, the magnetic field is changed and induces a voltage in the coil. When a wheel with teeth is rotated in front of the pickup, the pickup output voltage rises each time a tooth passes by. Reference 4 lists the relative advantages and disadvantages of a proximitor probe when compared to a magnetic pickup for speed measurement as follows:

Advantages of Proximitor Probes

- The distance between the proximitor probe tip and the gear tooth can be larger, which reduces the chance of the tooth hitting (wiping out) the sensor due to radial vibration of the shaft.
- The output amplitude of the proximitor probe does not decrease or drop out at low speeds. (Rare earth metals can be used for the magnet in a magnetic pickup to increase its output amplitude and extend its range to lower speeds.)
- The proximitor probe is designed to be intrinsically safe for use in electrically hazardous areas.
- The wheel teeth need not be magnetic.
- The proximitor probe can be gapped blind (a feeler gage is not needed during installation to measure the distance of the probe tip from the teeth) since the probe senses the distance of a stationary surface.
- The proximitor probe output impedance is lower so that there are fewer noise and loading problems.

Disadvantages of Proximitor Probes

- The RF excitation of the proximitor requires a power source.
- The proximitor probe tip width is limited to 0.125 inch. This limits the number of teeth on the wheel and the measurement resolution since the tooth width cannot be smaller than the tip width.

Speed measurement resolution error can cause poor speed and throughput control. A dynamic simulation of a compressor by Monsanto showed that the speed measurement resolution had to be better than 10 rpm for that application in order to maintain controllability. The following equation can be used to determine the number of teeth and the digital tachometer sample time required to meet a resolution specification:

$$n = 60/(T_s \cdot E_m) \quad (7\text{-}3)$$

$$T_s < E_o/(\Delta N/\Delta T) \quad (7\text{-}4)$$

where:

E_o = overspeed error allowable

E_m = measurement resolution error

n = number of teeth per revolution

$\Delta N/\Delta T$ = maximum rate of speed increase

T_s = digital tachometer sample and hold time

The above equation shows that decreasing the sample and hold time or increasing the number of teeth per revolution increases the speed measurement resolution. The minimum resolution of digital tachometers is about 1 rpm, whereas the minimum resolution of analog tachometers is about 10 rpm (Ref. 28).

The basic circuit components are shown in Figure 7-2a for an analog tachometer and in Figure 7-2b for a digital tachometer. The output of the pickup is digital but is converted to a continuous voltage output by a three-stage RC filter network in the analog tachometer. Component accuracy and temperature stability limit the analog measurement accuracy. In a digital tachometer, the pickup digital pulses are counted and compared to a reference count from an accurate oscillator. The register-comparator generates an interrupt for the computer to read the count. The computer resets the counters and holds the count over the sample interval.

Key Concepts

- Speed sensors are usually proximitor probes or magnetic pickups.
- Proximitor probes are less likely to be wiped out by vibration.
- Proximitor probes can measure down to and including zero speed.
- Magnetic pickups permit more teeth than proximitor probes.
- Measurement resolution depends on the number of teeth and the sample time.
- Digital tachometers can achieve greater measurement resolution.

7-4 FLOW MEASUREMENT

The flow measurement should be as fast as possible. Most electronic d/p transmitters have a time constant adjustable from about 0.2 to 1.7 seconds. A pneumatic

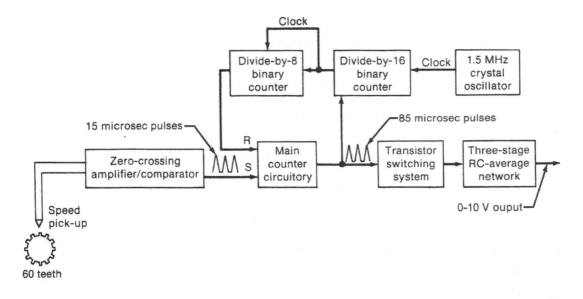

Figure 7-2a   Analog Tachometer Circuitry *(Courtesy Westinghouse)*

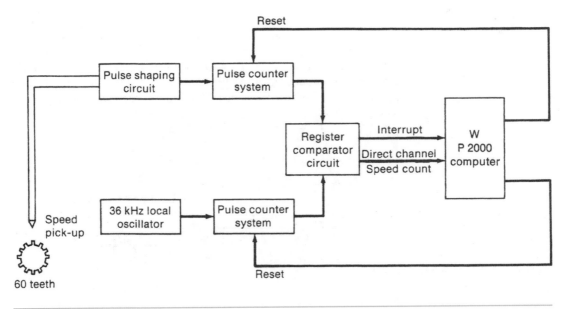

Figure 7-2b   Digital Tachometer Circuitry *(Courtesy Westinghouse)*

d/p transmitter with damping can have a time constant as large as 16 seconds. A diffused silicon electronic d/p transmitter can have a time constant as small as 0.05 second.

Figure 7-3 shows how well transmitters with different time constants reproduce the initial precipitous drop and subsequent oscillations in flow associated with surge. The operator and the control system would not even realize that surge was occurring if a pneumatic transmitter with damping or long signal lines was being used. An electronic transmitter with its damping at maximum (1.7 sec) is almost as bad. An electronic transmitter with its damping at its minimum (0.2 sec) would miss the initial drop in flow but would produce surge oscillations with a delay of 0.2 second and about 60 percent of the amplitude of the real oscillations. The diffused silicon transmitter accurately reproduces the initial precipitous drop and subsequent oscillations in flow.

The most common mistake made in the design of the surge flow measurement primary is the installation of the primary without sufficient upstream and downstream straight runs. If the duct or pipe is large, the straight runs are expensive and use up extensive space in the equipment layout on the plot plan drawing. Moreover, the customer frequently is not adequately warned about the inaccuracies and the noise that result from insufficient straight runs. A noisy flow measurement will require that the controller gain be decreased, which will increase the required distance from the surge curve of the suction flow controller set point. If

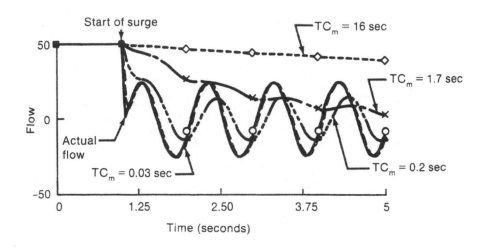

Figure 7-3   Effect of Measurement Time Constant on Flow Drop and Oscillations of Surge

the cost of the straight runs is prohibitive, the best alternative is a set of straight-ening vanes and a flow tube. A supersonic flow nozzle does not require straight runs downstream because the pressure waves propagate at the speed of sound, whereas the fluid is traveling at a velocity greater than the speed of sound. Test data shows that the use of a high-pressure tap upstream of a streamlined flow tube with a piezometric ring for the low-pressure tap minimizes the phase shift of these pressure waves from upstream or downstream discontinuities. This method also yields the greatest cancellation of the noise by simultaneous appearance at the high- and low-pressure ports of the d/p transmitter (Ref. 7).

Key Concepts

- Most pneumatic transmitters will not detect surge at all.
- Most electronic transmitters will not detect the precipitous flow drop.
- Diffused silicon transmitters will detect the precipitous flow drop.
- Flow primaries are often installed with insufficient straight runs.
- Straightening vanes and cancellation can reduce measurement noise.
- Measurement noise increases the required surge controller set point.

7-5 TEMPERATURE MEASUREMENT

Temperature measurements are used for bearing and oil high-temperature inter-locks, surge detection, temperature correction of discharge flow measurements, and compressor efficiency calculations. Temperature measurement error is not important for the interlocks since the temperature difference between normal operation and shutdown is relatively large. The temperature measurement time constants create a dynamic temperature error that is important for surge detec-tion and temperature correction of discharge flow measurements. The amplitude of measured temperature oscillations versus actual temperature oscillations can be estimated with the following equation (Ref. 21).

$$\text{For } \tau_m > T_o/2$$

$$A_m = (A_p \cdot T_o)/(2\pi \cdot \tau_m) \quad (7\text{-}5)$$

where:

A_m = amplitude of the measured oscillation (percent)

A_p = amplitude of the actual process oscillation (percent)

T_o = period of oscillation (sec)

τ_m = time constant of the measurement (sec)

The temperature response of a sensor in a thermowell can be approximated by a large and a small time constant in series. These time constants depend on the gas velocity, the annular clearance in the thermowell, the annular fill, and the sensor construction. A thermocouple in a thermowell with 0.04-inch annular clearance of air will have a large time constant that varies from 107 seconds at 5 fps to 92 seconds at 300 fps. A 100-second measurement time constant will cause the amplitude of the measured oscillations during surge to be less than 0.2 percent of the actual temperature oscillations. The time constant can be reduced to less than a second by the use of a bare thermocouple element with an exposed loop (sensing junction is in direct contact with the gas). A I-second measurement time constant will cause the amplitude of the measured oscillations during surge to be about 20 percent of the actual temperature oscillations. If the gas temperature is not oscillating but is ramping, the measured temperature will track the gas temperature with a delay equal to the sum of the large and small measurement time constants. For example, if the measurement time constants are 100 seconds and 10 seconds, the measured temperature will reach the gas temperature 110 seconds or 110 surge cycles later. Measurement attenuation of oscillations and measurement delay of transitions in gas temperature are severe problems for both surge detection and discharge flow measurement correction. A technique called the loop current step response (LCSR) method can accurately calculate the time constant of the temperature sensor in service by (1) applying a small current to the sensor leads to internally heat the sensing element and (2) analyzing the resulting temperature transient (Ref. 19). The temperature measurement response can be improved for control systems by the use of a lead-lag station to apply a lead time equal to the calculated time constant of the sensor.

Dynamic temperature measurement errors are not a problem for compressor efficiency measurements since these measurements are not used for protection or control. If the measurements and calculations are made manually, they should be made when the operating point is relatively fixed. If the efficiency is calculated by a computer on-line and trended, the other measurements used in the calculation, such as flow and pressure, can be filtered by the computer with time constants equivalent to the temperature measurement time constants in order to avoid inaccurate trends during load changes.

Temperature measurements in high-velocity gas can cause the measured temperature to be greater than the gas temperature due to the conversion of velocity energy to thermal energy upon gas impact on the sensor. For a gas at 950° F and Mach 0.7, the measured temperature would be about 125 degrees higher than the gas static temperature. A properly designed shield around the temperature sensor could reduce this error to less than 4 degrees. The random component of this aerodynamic recovery error is only about 0.3 percent (Ref. 9).

Key Concepts

- Sensors in thermowells are too slow for surge detection or control.
- Sensors with an exposed junction are much faster.
- Sensors in high velocity will read too high unless a shield is used.

7-6 VIBRATION AND THRUST MEASUREMENTS

Vibration and thrust measurements are important for startup (e.g., alignment and balance), maintenance (e.g., bearing wear), and protection (e.g., surge and bearing failure). A compressor will typically have two radial vibration measurements about 90 degrees apart called X and Y at the high- and low-pressure ends of the shaft. A key phasor, which consists of a proximitor probe to measure the once-per-revolution passing of a notch on the shaft, is used to measure the shaft speed and provide a reference for the phase shift of the vibration oscillations. The axial thrust of the rotor at the driver end is also measured by the use of redundant proximitor probes. Figure 7-4 shows a typical system layout of these measurements for a centrifugal compressor.

Proximitor probes are typically used for measurement of shaft radial vibration and axial thrust. Figure 7-5 shows the input and output signals of a proximitor. The proximitor supplies an RF signal to the probe. The probe modulates the amplitude and frequency of the envelope of the RF oscillations depending on the distance of the surface from the probe tip. As the surface comes closer to the probe tip, eddy currents are set up on the probe surface, power is absorbed, and the amplitude of the RF envelope is decreased. If the position of the surface

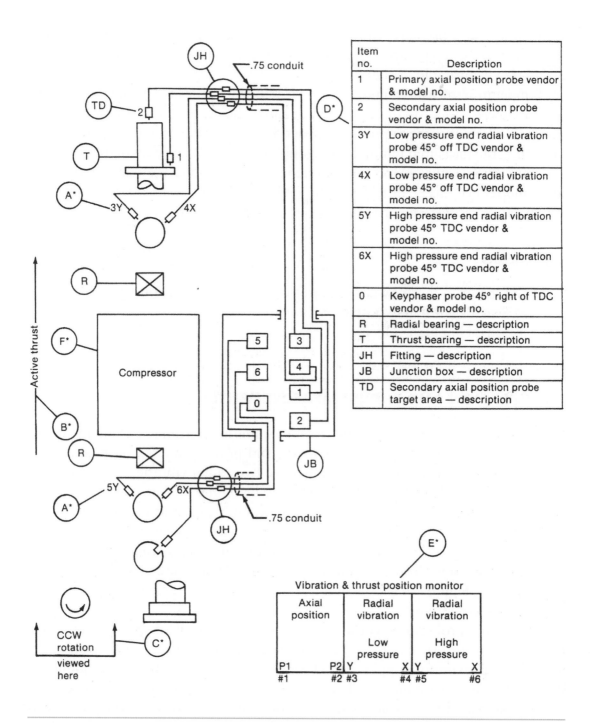

Figure 7-4   Typical Vibration System for a Centrifugal Compressor (API Standard 670)
(Courtesy American Petroleum Institute)

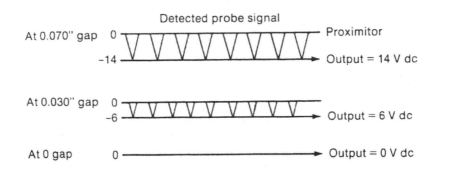

Figure 7-5   Proximitor Input and Output Signals
(Courtesy Bently Nevada Corporation)

is oscillating, the envelope is modulated with the same frequency as the surface. The proximitor output consists of the negative voltage peaks of the envelope. This output can be directly measured on a voltmeter. The proximitor plus probe is a gap-to-voltage transducer.

The output voltage from the proximitor is plotted versus mils (thousandths of an inch) of gap to generate a calibration curve. Figure 7-6 shows a calibration curve for an axial thrust proximitor and probe. The curve does not show that the output

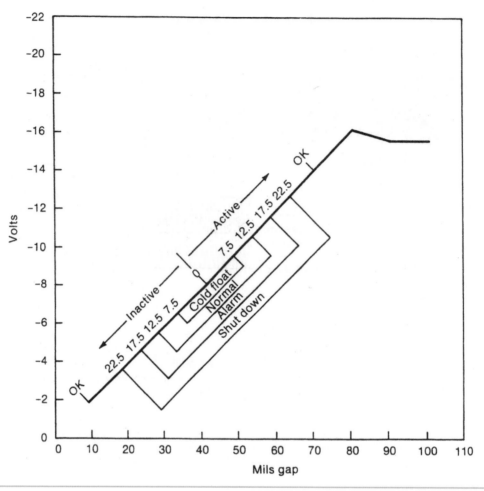

Figure 7-6   Typical Calibration Curve for Thrust Measurement (API Standard 670)
                    (Courtesy American Petroleum Institute)

voltage actually falls to zero somewhere below 10 mils (0.010 inch). Also, the output voltage levels off above 80 mils. The usable linear span of voltage versus gap is about 70 mils. The calibration curve will shift to the left as the conductivity of the surface increases. It will shift to the right as the temperature at the probe increases. The linear span of voltage decreases as the tip diameter decreases (Ref. 4). It is more important that the linear span be large for thrust measurements rather than for vibration measurements, because the change in axial position is much greater than the amplitude of the radial vibration of the shaft. Figure 7-6 shows that the normal range of thrust movement is near the middle of the linear range and that there are alarm and shutdown settings on both sides of the operating range. The axial movement may normally be in the direction of the normal or active thrust pads, but a sticky coupling, surge, or worn balance seals can cause thrust in the opposite direction to the abnormal or inactive thrust pads. A recording of thrust measurement will show a staircase for shaft movement through the float zone or sharp spikes for a shaft against a thrust bearing for a sticky coupling. A recording for surge will show continuous shaft movement in both directions (Ref. 4).

Since the vibration measurement uses a gap-to-voltage transducer, any eccentricity of or mechanical blemish on the shaft will show up as radial vibration. This measurement noise is called runout. The vibration monitors are designed to suppress from 0.1 to 0.5 mil of runout. Runout can be distinguished from vibration by recognizing the following characteristics of runout (Ref. 4):

- Runout due to mechanical blemishes is non-sinusoidal.
- Runout pattern does not change with speed.
- Runout pattern at X probe will appear 90 degrees later at Y probe.

The amplitude of casing vibration measurement may only be about 0.5 mil while normal noise may also be 0.5 mil, so that the signal-to-noise ratio is not as good as that for shaft vibration measurement.

Since proximitors and probes are not as commonly used as many other instruments and since the gap clearances being sensed are relatively small, there are special installation requirements. Some of the more frequent installation pitfalls are (Ref. 4):

- Probes mounted over chrome plating, hubs, and sweat-on collars.
- Radial probes mounted at angle from perpendicular to shaft centerline.
- Axial probes mounted at end opposite from thrust bearing.*
- Probes do not extend through the mounting bracket or housing.
- Probe mounting brackets flex and resonate.
- Excessive strain of conduit connection to small probe body.
- Probes gapped at 25 mils since linear span was said to be 50 mils.
- Proximitor housing filled with oil due to poor oil seal in conduit.

---

* Axial probes have been deliberately mounted at the opposite end as well to measure the axial clearance of the final stage, which is important when all the thermal clearances are used up due to internal reheating of the gas during surge oscillations.

- Probe cable broken from screwing in probe with cable attached.
- Different probe cable length used for calibration and installation.
- Proximitor output cable run parallel to power lines.
- Steam condensation in proximitor housing.

The shutdown levels for the axial thrust should be set just far enough away from normal operation to show evidence of movement into the thrust bearing babbit upon inspection; otherwise, the instrument may be blamed for a false shutdown. The shutdown level must not be set so far away that the babbit is removed and the rotor makes contact with the bearing (Ref. 18). Thrust bearing force and temperature sensors will show an increase in force and temperature as the rotor moves into the babbit. However, after the babbit is removed and axial rub starts, the force and temperature drop back down, signaling a false all clear condition. The proximitor probe will continue to indicate increasing axial displacement after babbit removal. If the rotor moves far enough to wipe out the probe tip, the proximitor will go to zero, which indicates high thrust (Ref. 4). A 2-out-of-2 voting instead of a 2-out-of-3 voting system for shutdown can be used to eliminate false trips because probe, transducer, or cable failure will indicate a high thrust condition.

Key Concepts

- Vibration and thrust measurements are used for startup, maintenance, and shutdown.
- Proximitors are typically used for vibration and thrust measurement.
- Proximitors are gap-to-voltage transducers.
- Surface conductivity and probe tip temperature affect the calibration.
- There are many installation pitfalls.
- Shaft eccentricity and blemishes will cause noise called runout.
- Thrust measurements indicate high thrust for a loss-of-signal failure.
- A 2-out-of-2 voting system is used for thrust shutdown.

REFERENCES

McMillan, G.K., *Process Control Case Histories*: *An Insightful and Humorous Perspective from the Control Room*, "Compressor Surge Control: Traveling in the Fast Lane," Momentum Press, 2010

McMillan, G.K., *Essentials of Modern Measurements and Final Elements in the Process Industry—A Guide to Design, Configuration, Installation, and Maintenance*, ISA, 2010.

QUESTIONS

1. Why is anti-reset windup needed in the surge controller?

    ANSWER

2. Why must a surge control valve throttle quickly?

    ANSWER

3. How do you make a large control valve stroke faster?

    ANSWER

4. Why is a bypass needed on a booster on the output of a positioner?

   ANSWER

5. How can speed measurement resolution be increased?

   ANSWER

6. What is the most common installation problem with flow primaries?

   ANSWER

7. Why can't a sensor in a thermowell be used as a surge detector?

   ANSWER

8. Why aren't magnetic pickups used for vibration or thrust measurements?

   ANSWER

Disturbances

8-1 THROUGHPUT CONTROL

Any change in operating conditions that causes the operating point to move on the compressor map is a disturbance. Some sources of throughput control disturbances are:

- Suction valve hysteresis.
- Vane position hysteresis.
- Turbine steam control valve hysteresis.
- Speed control regulation or variation error.
- Change in demand of downstream user.
- Fouling of downstream equipment.
- Blowing of downstream relief valve or rupture disc.
- Change in pressure set point of downstream user.
- Change in molecular weight or temperature of suction gas.
- Change in compressor efficiency due to failure or wear of seals.

The peak error for a given disturbance increases as the controller gain decreases, the process gain increases, and the time constant of the disturbance decreases. The allowable controller gain decreases as measurement time constants, actuation time constants, and measurement noise increase. Thus many of the same instrument requirements for good surge control also apply for tight throughput control. The process gain increases the open-loop error (error with controller on manual) for any given disturbance. The process gain is determined from the compressor map (see Section 5-4 for method) and can be reduced by proper pairing of the manipulated and controlled variables. The time constant of the disturbance can be increased by slowing down the stroke of the user's flow and pressure control valves. Also, since it is actually the magnitude of the disturbance time constant relative to the throughput control loop period that is important, the disturbance can be made to seem slower to the throughput control loop by making the loop faster through the use of faster instruments. The following equation can be used to estimate the peak error (Ref. 21):τ

$$E_x = \{1/(K_c \cdot K_o)\} \cdot E_o \cdot e^{\{(-4 \cdot \tau_d)/T_u\}} \quad (8\text{-}1)$$

where:

E_x = peak error (flow or pressure units)

K_c = controller gain (dimensionless)

K_o = open-loop gain (%/%)*

E_o = open-loop error (flow or pressure units)

τ_d = time constant of the disturbance (seconds)

T_u = control loop period (seconds)

Key Concepts

- Anything that moves the operating point is a throughput disturbance.
- The major sources of disturbances are actuation and users.
- The peak throughput error increases as the controller gain decreases.
- The peak throughput error increases as the process gain increases.
- The peak throughput error increases as the disturbance gets faster.

---

* The open-loop gain can be determined by putting the throughput controller on manual, increasing the controller output, and dividing the percent change in controller measurement by the percent change in controller output. The minimum allowablecontroller gain is inversely proportional to the open-loop gain (see Section 12-1).

8-2 SURGE CONTROL

The surge control loop is different from other types of control loops because it is desirable that this loop keep the measurement above and not at the set point. It is desirable to keep the operating point as far to the right of the surge curve as allowable through user process constraints and operating efficiency. The surge controller set point is more a lower limit than a control point for suction flow. The peak error (magnitude of excursion past set point towards the surge curve) can be estimated by use of Equation 8-1 for surge disturbances. The throughput control disturbances listed in Section 8-1 become surge control disturbances if they cause the operating point to approach near enough to the surge set point to cause the surge control valves to open.

It is obvious that the surge control valve time constant must be smaller than the disturbance time constant. Thus if the disturbance is caused by a downstream block or check valve slamming shut, the stroking time of the control valve must be faster. In one application within Monsanto, a hydraulically operated check valve downstream would slam shut in a fraction of a second when tripped by low hydraulic pressure. The result was severe uncontrollable surge and extensive thrust bearing damage. The solution was to de-energize a solenoid-actuated ball valve to exhaust the surge valve actuator pressure when it was triggered by a low hydraulic pressure switch. It is important that the fastest possible disturbance be estimated before the surge control valve is ordered. Frequently the fastest disturbance is due to an equipment or instrument failure.

The disturbance can also be too slow. If the operating point approaches the surge controller set point slowly, the reset action of the controller integrates the error enough to cause the controller output to return to its maximum value and close the surge control valves. A slow measurement can result in the surge measurement being at the surge set point with the surge valves shut. Since the pre-stroke dead time is greatest at the closed position of the surge valves, especially for eccentric disc valves, the surge valves may be too slow for any subsequent disturbance. For a linear approach of the surge measurement to the surge set point, the time (time for error to go from initial error to zero error on the surge controller) must be less than twice the integral time setting (T_i) to prevent closing of the surge valves. For example, if the reset setting were 10 repeats per minute ($T_i = 6$ seconds per repeat), a disturbance time of 12 seconds would keep the surge control valves closed until after the surge measurement crossed set point even though the surge controller had anti-reset windup. The installed flow characteristic of a butterfly block valve will give a slower approach to set point and then a faster approach to the surge curve than will a linear block valve (see Appendix C, Figure C-4, for an example of the change in stroking time). A mass balance override or "anticipator" control system (described in Section 6-5) would open the surge valve before the operating point reached the surge set point if the lag time or velocity limiter were set to be slower than the closing block valve.

For a linear block and surge valve:

$$T_v < T_b > 2 \cdot T_i \quad \text{(8-2)}$$

where:

T_b = stroking time of the block valve to zero error (sec)

T_v = stroking time of the surge valve to prevent surge (sec)

T_i = integral time of surge controller (sec/repeat)

Key Concepts

- The surge control valve must be faster than the fastest disturbance.
- A disturbance too fast or slow can increase the peak error.

QUESTIONS

1. Why would throughput measurement noise increase the throughput control peak error?

   ANSWER

2. Would a quick-opening or a linear installed flow characteristic of a down-stream block valve be better for surge control?

ANSWER

Throughput and Surge Control Interaction

9-1 SEVERITY

The throughput and surge control loops both try to position the operating point on the compressor map. The control action of each loop affects the other; thus the two loops interact. The question is, "How severe is the interaction?" Interaction can cause inverse response, high-frequency uncontrollable noise, or continuous oscillations. The severity depends upon the controller gains of the two controllers, the periods of the two loops, the relative gain between loops, and whether the control action is one of conflict or cooperation. The interaction is worst for equal controller gains, equal loop periods, negative relative gains, and control action conflict. The controller gain depends upon the process time constant as well as instrument time constants and dead times. The loop period depends primarily on the instrument dynamics alone. Since the process time constant is about the same for both loops, the controller gain and the loop period are about equal for the two loops if both use fast instruments. The throughput controller controller gain is purposely decreased in some applications to reduce interaction. However, decreasing the controller gain unfortunately increases the throughput control peak error. The pneumatic actuators for suction throttling, vane positioning, and turbine steam throttling are slower than the actuators for surge control valves. The slower actuation response means that the maximum allowable controller gain will be smaller and the loop period will be larger for the throughput controller. Thus the interaction is greater for throughput loops with hydraulic actuators or variable speed drives.

The *steady-state gains* for each pairing of controlled and manipulated variables can be calculated from the compressor map, the control valve capacities and flow characteristics, and the transmitter ranges. The change in percent measurement for a change in percent control valve position at a normal operating point is calculated for each pair and put in a square 2 × 2 matrix. The relative gain matrix is this steady-state gain matrix multiplied by the corresponding elements of the inverse of this matrix. The best pairing of measurements and control valves are those combinations whose relative gains are closest to unity. It is desirable that all other possible combinations have relative gains close to zero. A relative gain element between zero and one means that the net effect of the other loop is to change the measurement signal in the same direction (cooperation). A relative gain element greater than one means that the net effect of the other loop is to change the measurement signal in the opposite direction (conflict). A relative gain element less than zero means that the net effect is also conflict but the interaction dominates and causes the gain to change sign. Thus, if this loop is stable when the other loop is in manual, it probably is not stable when the loop is on automatic. Even if the loop is stable, the measurement will have to overshoot the set point to reach steady state (Ref. 21).

The operating point follows a path on the load curve for throughput control. It follows a path on the characteristic curve for surge control. The effect of the surge controller on a pressure controller is greatest for a steep characteristic curve. The effect of a pressure controller on the surge controller is greatest for a flat load curve (vice versa for flow throughput control). Figure 9-1 is a compressor map for an axial compressor and a system with a high static load. It illustrates the conditions for high relative gains and high interaction between surge and pressure control.

Conflict of control actions is particularly dangerous for surge control because the reaction of the throughput control loop is to drive the operating point closer to the surge curve when the surge control loop is trying to prevent a potential surge. Any excursion to the left of the surge curve, no matter how brief, will cause the precipitous drop in flow and start the surge cycles. Unfortunately, this interaction is not observed during normal operation when the surge valves are closed. The operators may not be aware of the interaction problem and therefore may have a false sense of security. Since many surge conditions develop rapidly and

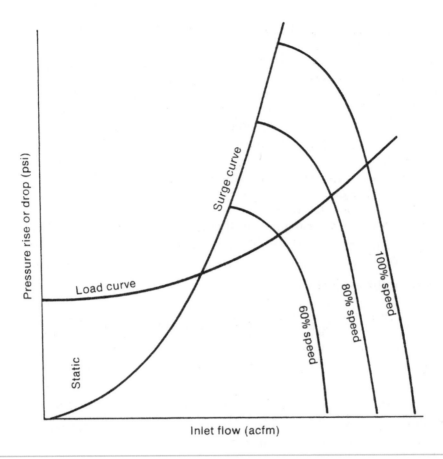

Figure 9-1   Characteristic Curves and Load Curve for High Interaction

shutdown occurs immediately for uncontrollable surge, there may not be sufficient time to diagnose the problem properly. Whether the throughput and surge
control loop actions are of conflict or cooperation depends on the relative slope
of the load curve compared to the surge curve and the pairing of the manipulated
and measured variables. The relative difference between the slopes determines in
which direction the surge and load curves are converging. If the surge curve bends
over (high compression ratio or adjustable guide vanes), the surge curve is steeper
at low pressures. The load curve is steeper at high pressures unless the load curve
is flat due to high static loads or parallel compressors. Table 9-1 summarizes the
results of different relative slopes and pairings of variables for throughput control.
The controlled variables are discharge pressure and flow. The manipulated variables are speed, guide vane position, and suction valve position.

Table 9-1   Effect of Relative Slopes and Pairing on Interaction

Pairing of variables	Load curve steeper	Surge curve steeper
Pressure & speed	Cooperation	Conflict
Pressure & vane	Cooperation	Conflict
Pressure & valve	Cooperation	Conflict
Flow & speed	Conflict	Cooperation
Flow & vane	Conflict	Cooperation
Flow & valve	Conflictw	Cooperation

Once a compressor goes into surge, the severity of the surge increases as the compressor speed increases (see Equation 3-5). Thus for the pairing of flow and speed in Table 9-1, the throughput controller will unfortunately try to greatly accelerate the compressor when the flow precipitously drops at the start of surge. If the response time of speed is small due to hydraulic actuators or tachometer feedback, the severity of the surge cycle is drastically increased.

Key Concepts

- Interaction is worst for equal controller gains and loop periods.
- Interaction is worst for relative gains greater than 1 or less than 0.
- The relative gain depends on the slopes of load and characteristic curves.
- The relative gain depends on the type of throughput control.
- Interaction is worst for conflict.
- Conflict depends upon the relative slopes of surge and load curves.
- Conflict depends upon the type of throughput control.

9-2 DECOUPLING

If the interaction is severe and sufficient detuning of the throughput controller is not possible, half decoupling should be used. Figure 9-2 shows the addition of a summer on the output of the slower throughput controller output to add the instantaneous output and subtract the lagged output of the surge controller. The resulting output signal from the summer is used to manipulate either the speed, the vane position, or the suction valve position. The signs at the summer depend

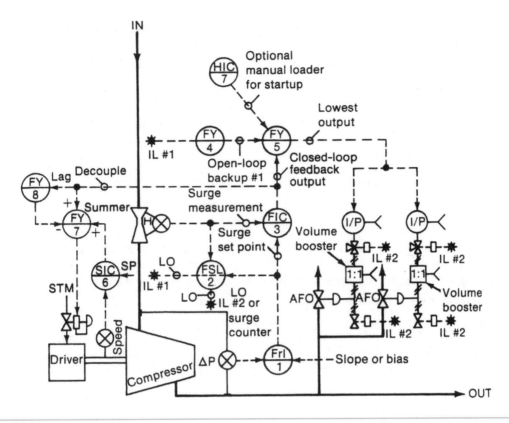

Figure 9-2   Half Decoupling the Throughput and Surge Controllers (A smart digital positioner with a booster and properly set bypass on the positioner output is used instead of an I/P and booster for reasons described that support Figure 7-1b)

on the failure position of the throughput control valve and the relative slopes of the load and surge curves. The use of a lagged output signal cancels the effect of the instantaneous signal for normal operation (surge valves closed). If the operating point approaches close enough to the surge controller set point, the surge controller output decreases, which then also decreases the speed, the vane position, or the suction valve position for a load curve steeper than the surge curve.

Key Concepts

- The decoupler is a summer on the throughput controller output.
- The decoupler does not affect normal operation.

QUESTIONS

1. Would a pneumatic throughput control loop require a decoupler?

    ANSWER

2. Why would a mass balance override or "anticipator" control system interact more often with the throughput control system?

    ANSWER

STUDENT SUMMARY NOTES AND QUESTIONS FOR INSTRUCTOR

Multiple Compressor Control

10-1 SERIES COMPRESSOR CONTROL

Compressor stages (same driver shaft) or compressors (separate driver shafts) are installed in series to increase compression ratio. The control systems shown in the literature use a single surge control valve on the end of the train with a low-signal selector (air-fail-open surge valve) to choose the output of the surge controller on the stage or compressor that is closest to surge. Separate surge control valves are not shown in order to avoid the interaction that might result. The opening of an intermediate surge valve would increase the flow to an upstream stage or compressor but would decrease the flow to a downstream stage or compressor. However, field experience shows that the interaction is minimal. The use of a single surge valve on the end of the train delays the correction of flow for the upstream stages or compressors, especially if there are intercoolers or separators between the stages or compressors. It is best to duplicate the surge control system for each stage or compressor in series. If the throughput flow is not equal for each compressor in series (caused by separators and recycle streams), then compression ratio dividing (per the following equation and shown in Figure 10-1) would help prevent unnecessary opening of the surge valve (Ref. 26).

$$K_1 \cdot (P_1/T_1) \cdot \Delta P_1 + B_1 = K_2 \cdot (P_2/T_2) \cdot \Delta P_2 + B_2 \quad (10\text{-}1)$$

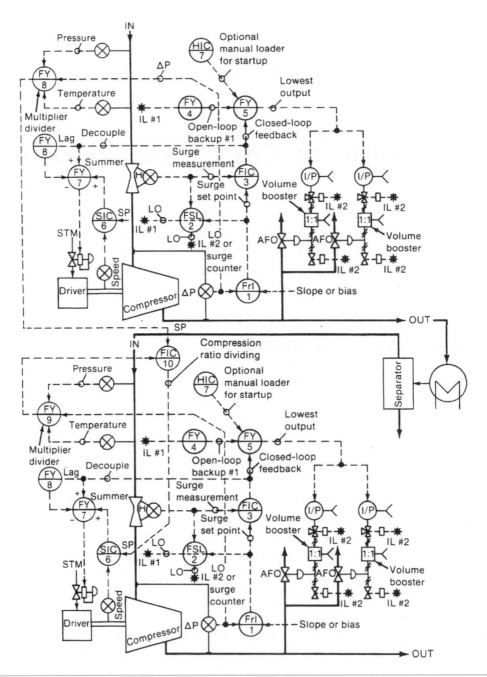

Figure 10-1 Compression Ratio Dividing for Compressors in Series (A smart digital positioner with a booster and properly set bypass on the positioner output is used instead of an I/P and booster for reasons described that support Figure 7-1b)

where:

B_1 = bias of surge set point of compressor #1

B_2 = bias of surge set point of compressor #2

K_1 = slope of surge set point of compressor #1

K_2 = slope of surge set point of compressor #2

P_1 = suction pressure of compressor #1

P_2 = suction pressure of compressor #2

ΔP_1 = pressure rise developed by compressor #1

ΔP_2 = pressure rise developed by compressor #2

T_1 = suction temperature of compressor #1

T_2 = suction temperature of compressor #2

Sometimes flow measurements are not available for the total flow between stages because of internal flows or unmeasurable recycle flows. If a partial flow can be measured, then an open-loop surge protection system that uses the derivative of this flow to pop open the surge valve for that stage is better than relying solely on the surge control for other stages.

Key Concepts

- Each stage or compressor in series should have a surge control system.
- Compression ratio dividing is used for compressors in series.
- Flow derivative backup should be used for partial flow measurements.

10-2 PARALLEL COMPRESSOR CONTROL

Compressors are installed in parallel to increase the flow capacity. Most parallel compressor installations to date use a graduated set point method of distributing load among the compressors. The compressor with the lowest set point is unloaded first, while the compressor with the highest set point is unloaded last. This method of distribution is inefficient for the following reasons: (1) the lower set point compressors may be operating with surge control valves partially or fully open for load reduction; and (2) the operating pressure will be higher since the incremented set points are typically discharge pressure. The incremented set points may also upset the downstream users. A better control strategy recognizes how far each compressor operating point is from the surge curve and simultaneously loads and unloads all compressors proportionally. The leader compressor is either the largest compressor or the compressor whose operating point is closest to the surge curve for a load increase or furthest from the surge curve for a load decrease (Ref. 26). If the bias divided by the slope of the surge set point curve is chosen to be equal for all compressors, then the ratio controller set point of each follower compressor is simply the flow of the leader compressor multiplied by the slope of the follower surge set point divided by the slope of the leader surge set point. Equation 10-2 shows this relationship and Figure 10-2 shows the arrangement of control loops for two compressors in parallel.

$$H_f = H_l \cdot (K_k/K_l) \quad (10\text{-}2)$$

where:

H_f = suction flowmeter differential head of follower compressor

H_l = suction flowmeter differential head of leader compressor

K_f = surge set point slope for follower compressor

K_l = surge set point slope for leader compressor

Key Concepts

- Leader selection and load dividing saves energy.
- Leader selection and load dividing reduces upsets to users.

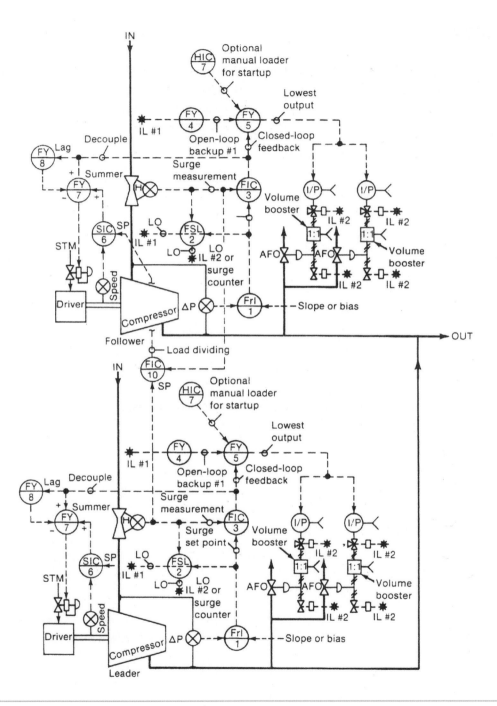

Figure 10-2   Load Dividing for Compressors in Parallel (A smart digital positioner with a booster and properly set bypass on the positioner output is used instead of an I/P and booster for reasons described that support Figure 7-1b)

QUESTIONS

1. Why does the need for individual surge valves for compressors in series increase as the size of an intervening separator increases?

   ANSWER

2. Why is the leader for parallel compressor control chosen on the basis of operating point proximity to the surge curve?

   ANSWER

Computer Monitoring

11-1 FREQUENCY ANALYSIS

The signals from the radial vibration and key phasor contain a wealth of information on the health of the compressor. The vibration monitors in the control room utilize only a small part of its quantity and realize only a small part of its potential. The quantity of data is so great and the methods so complex that it is an ideal application for a computer. Computers have been successfully applied to a number of large compressor installations. Some of the economic incentives are (Ref. 2):

- Longer runs between shutdowns due to elimination of many unscheduled shutdowns by corrective action before problem progresses.
- Shorter shutdowns because unnecessary maintenance is avoided and the material and manpower for required maintenance can be arranged for prior to shutdown.
- Scheduling shutdowns at optimum time by early warning of required maintenance.
- Reduced number of catastrophic shutdowns, which is important not only for preventing extensive machinery repair and lost production costs but also for personnel safety.

The computer can be designed to provide the following special displays in addition to the typical trend plots versus time (Ref. 17):

- Dynamic shaft orbits
- Bode plots
- Projected trends
- Time correlation of data
- Signature analysis
- Spectral map

Dynamic shaft orbits show the path traced by the centerline of the shaft relative to the casing by plotting the X radial proximitor output against the Y radial proximitor output. The X and Y inputs of an oscilloscope are used to generate the display of the orbit for balancing. Many major machinery problems can be diagnosed from the shape of the orbit. Bode plots are plots of vibration amplitude and phase versus speed. The burden placed on the computer and the programmers can be considerably reduced by using a vector filter phase meter to determine the phase and amplitude and present it in a compatible digital form to the computer. Projected trends show the future path of the vibration amplitude at a frequency of interest based on the present rate of change. This allows a problem to be corrected before the amplitude reaches the shutdown level. Time correlation of data presents all events that occurred at the same time that a machinery problem started. Signature analysis compares the present plot of vibration amplitude versus frequency at a speed with a reference plot for deviations. The change in vibration amplitude at certain frequencies and speeds can be used to diagnose the problem. The computer must have an array processor fast enough to do the computations for the Fast Fourier Transform (FFT) for spectrum display (an array processor does math calculations in parallel rather than in series). The FFT converts the information from the time domain into the frequency domain. A real time analyzer (RTA) can be used to compute the FFT so that a standard process control computer can be used. A spectral map is a three-dimensional plot of vibration amplitude, frequency, and speed. This map is used to determine which compressor speeds contain the most information in terms of vibration amplitude versus frequency for signature analysis.

Key Concepts

- Computer frequency analysis increases compressor on-stream time.
- Computer frequency analysis reduces compressor repair costs.
- A real time analyzer (RTA) permits the use of a standard computer.

11-2 COMPRESSOR MAPPING

The computer can display on a CRT a compressor map that shows the efficiency ellipses, the surge set point, and the surge, load, choke, and characteristic curves. This display is useful for monitoring slow approaches to surge and changes in operating efficiency. A slow approach to the surge curve may be halted by adjusting the user demand or the compressor load distribution for parallel compressors. If the power input is measured, the operating efficiency can be calculated and compared to the ellipse reference value to detect the start of seal leaks. If the effect of operating conditions on the surge curve can be estimated by an equation, the surge curve can be updated on-line. The computer can record surge points from deliberate surge tests or accidental surges if the suction flow and pressure rise transmitter time constants and the computer sample time are less than 0.05 second. A flow derivative calculation is made to signal the start of surge and to save the previous record of flow and pressure rise to be used for the coordinates of the surge point. In one Monsanto application, the compressor accelerates so much at the start of surge that the derivative of the speed is used to signal the start of surge. A compressor map display of the surge curve is particularly useful for high-compression ratios to show the surge curve breakpoints and the intersection point with the choke (stonewall) curve. The accuracy of a characterized surge set point can be monitored and the required offset of the set point from the surge curve verified. The computer can be used to trend the actual path of the operating point on the compressor map during a surge cycle and to provide a plot of the negative flow branch of the characteristic curve.

Key Concepts

- Compressor mapping can increase operating efficiency.
- Compressor mapping can halt slow approaches to the surge curve.
- A fast computer with fast measurements can record surge points.

QUESTIONS

1. Why does the use of a real time analyzer permit the use of a standard process control computer?

   ANSWER

2. Why do the transmitter time constants and computer sample time have to be less than 0.05 second to record surge points on-line?

   ANSWER

STUDENT SUMMARY NOTES AND QUESTIONS FOR INSTRUCTOR

Startup

12-1 TUNING

The controller mode settings depend on the speed of the instruments and the amount of measurement noise. The ultimate oscillation method of controller tuning cannot be used because the control loop oscillations may trigger the start of surge oscillations, which have no relationship to the loop period. The loop period is approximately equal to four times the summation of all the instrument time constants and dead times. The integral time is set equal to 80 percent of the loop period. The controller gain is inversely proportional to the loop period (Ref. 21).

$$T_u = 4 \cdot (\theta_v + \theta_m + \tau_m + \theta_c + \tau_c) \quad (12\text{-}1)$$

$$\tau_c = 0.6 \cdot T_s \ (0.03 \text{ second for analog controller}) \quad (12\text{-}2)$$

$$\theta_c = 0.5 \cdot T_s \quad (12\text{-}3)$$

$$T_i = 0.8 \cdot T_u \quad (12\text{-}4)$$

$$\tau_p = 60 \cdot \{(V \cdot P)/(P_a \cdot F)\} \quad (12\text{-}5)$$

$$K_c = (2\pi \cdot \tau_p)/\{(4 \cdot T_u) \cdot K_o\} \quad (12\text{-}6)$$

where:

K_c = controller (dimensionless)

K_o = open-loop gain (percent/percent)

F = flow through the control valve (acfm)

P = pressure upstream of control valve (psia)

P_a = atmospheric pressure (psia)

τ_c = time constant of controller measurement filter (sec)

τ_m = time constant of transmitter (sec)

τ_p = time constant of pressure response (sec)

τ_v = time constant of control valve (sec)*

θ_c = dead time of digital controller (sec)

θ_m = dead time of transmitter (sec)

θ_v = prestroke dead time of control valve (sec)**

T_i = integral time setting (sec/repeat)

T_s = sample time of digital controller (sec)

T_u = ultimate period of the control loop (sec)

V = volume of the piping and equipment (cuft)

The controller gain is approximately 2 and the reset setting is approximately 20 repeats/min for a well-designed surge control loop. Many surge loops in service have controller gains smaller than this due to measurement noise and slow instruments, and to the rule of thumb that flow controllers use acontroller gain of 0.4 or less. The surge controller settings can be tested by increasing the bias setting of the ratio station to open the surge valves and by noting the closed response of the surge controller. The throughput controller should also be on automatic to check for interaction problems. The overshoot of the surge controller must be less than the distance of the backup set point from the feedback set point. Both the surge and the throughput controllers must be placed in manual to record the open-loop gain or pressure time constant or to use the reaction curve method of controller tuning (the manual output of each controller is changed by 10 percent and the resulting change in both measurements is trended on a high-speed recorder). The steady-state gain matrix can be completed from the trend recording data.

Key Concepts

- The ultimate oscillation method cannot be used for tuning.
- Both controllers must be on manual for the reaction curve method.
- The controller mode settings improve for faster instruments.

---

* τ_v can be approximated as ¼ the stroking time to the required position.

** θ_v must be increased to include the effect of stroke and booster dead band.

12-2 COMMISSIONING

The surge control system should be started up with the feedback surge controller, the throughput controller, and the signal generator on automatic. The feedback surge controller should be on remote set point. For surge testing, the bias of the ratio station can be decreased to move the surge set point temporarily just to the left of the surge curve. The ratio station should have an output limit to protect against improper slope or bias adjustments or pressure rise transmitter failure. If the suction flow is below this limit during startup, the surge valves will stay wide open. When the operating point exceeds this limit, the surge valves will throttle towards the closed position until the operating point reaches the surge set point. When sufficient forward flow is established, the surge control valves will close completely. Some operators refuse to relinquish manual control of the surge valves during startup or during surge testing. However, these are the times when the surge control system is needed most. Since the surge controller uses a remote set point and has an anti-reset windup kit, there is no need for the operator to interface with the surge controller or the associated stations during startup, normal operation, or shutdown. The surge controller should be left permanently on remote set point. (Many operators get confused about balancing the ratio station output and surge set point when switching to remote set point. It is best to avoid this problem by leaving the controller on remote set point.) If the operator uses a manual loader or an override controller's manual output to keep the surge valves open during startup, the surge valves will not close until all such signals to the low selector are increased to 100 percent.

Key Concepts

- The surge controller and stations should never be switched to manual.
- The surge controller should never be switched to local set point.
- The ratio station should have an output limit for protection.
- All override signals must be at 100 percent for the surge valves to close.

QUESTIONS

1. Which instrument typically affects the mode settings the most?

   ANSWER

2. What happens if the operator puts the surge controller on local set point?

   ANSWER

STUDENT SUMMARY NOTES AND QUESTIONS FOR INSTRUCTOR

Answers to Questions

Section 2

1. The characteristic curve for a reciprocating compressor is essentially vertical. The closing of a discharge valve increases the discharge pressure but does not decrease the flow. If the valve is closed too much, relief valves or rupture discs may blow open between the discharge valve and the compressor.

2. The characteristic curve for a centrifugal compressor is flatter. Changes in the load curve position due to changes in the pressure drop in the piping system will not cause as much a change in the discharge pressure.

3. While either centrifugal or axial compressors are suited for the stated flow and clean air service, the axial compressor has a greater operating efficiency so that the energy consumption is less (fewer kilowatt hours or pounds per hour of steam per pound of gas compressed).

Section 3

1. The intersection is at the point of zero slope (horizontal portion) of the compressor curve.

2. Both surge and stall occur only when the operating point crosses to the left of the surge curve.

3. The compressor flow drops precipitously to its minimum in less than 0.05 second.

4. The surge oscillations are so fast that the control loop reaction is too late. The resulting correction oscillation may be in phase with subsequent surge oscillations.

5. The amplitude and period increase at a rate proportional to the square root of the volume. The shape becomes more nonsinusoidal.

6. The amplitude increases at a rate proportional to the speed and the period increases slightly. The shape becomes more nonsinusoidal.

7. Damage due to high vibration, thrust, or temperature and reduced efficiency due to reduced clearances.

8. The time delays or time constants of instruments result in detection of surge after it has occurred because surge starts so quickly and its period is so small.

Section 4

1. The percent changes in atmospheric pressure expressed in psia are less than the percent changes in atmospheric temperature expressed in degrees R unless the suction duct is extremely long.

2. The surge set point would shift with temperature because the temperature in the flowmeter equation is different from the temperature in the surge curve equation.

3. The surge curve will bend over more because the ΔP of Equation 4-1 is proportional to the square root of the flow, while the h of Equation 4-4 is proportional to the square of the flow.

4. Adjustable guide vanes change the flow capacity by changing the suction pressure and gas prerotation.

Section 5

1. The throughput control methods progressing from least efficient to most efficient are discharge throttling, suction throttling, guide vane positioning, and speed control.

2. The throughput control methods progressing from least turndown to most turndown are discharge throttling, suction throttling, speed control, and guide vane positioning.
3. Throughput control errors are equal to speed control errors multiplied by the process gain.
4. The speed control loop response time determines whether an error is corrected before it causes a throughput control error and whether the throughput controller has to be detuned in the cascade control system.
5. A large filter time constant, large sample time, large pneumatic actuator, or undersized driver will cause a slow-speed loop time response.

Section 6

1. The minimum flow surge control system wastes energy at low pressures and provides reduced surge protection at high pressures.
2. The maximum-pressure surge control system wastes energy at high pressures and provides no surge protection at low pressures.
3. The flowmeter differential head is proportional to the square of the suction flow, but it needs to be proportional to the suction flow so that a linear ratio controller can be used.
4. The required offset of the set point from the surge curve increases as the disturbance time decreases and the controller gain decreases and integral times increases. The maximum controller gain and minimum integral time increases as the response time of the instruments in the loop increases. The addition of an open-loop backup system allows the offset to be reduced if overcorrection is not a problem.
5. The low signal selector ensures that the override control will not prevent the surge control system from protecting the compressor.
6. The downscale failure of any control system output will cause the surge valves to stroke wide open.

Section 7

1. Since the surge control valve is normally closed and air to close, the surge controller output is normally saturated and the surge valve

will not open until the operating point has crossed set point and is dangerously close to the surge curve unless the controller has anti-reset windup.

2. A slow throttling control valve will decrease the controller gain and increase the integral time settings, which will increase the distance required between the set point and the surge curve.

3. A large valve is made to stroke faster by the addition of a volume booster and the elimination of any restrictions on air flow into the actuator.

4. The combination of a positioner and a booster in series is unstable since the booster inlet volume is much less than the actuator volume.

5. The number of teeth on the wheel should be increased and the sample time of the digital tachometer should be decreased.

6. The most common problem is insufficient upstream and downstream straight lengths to produce an accurate and noise-free differential head signal for surge control.

7. The large measurement time constant would attenuate the oscillations in temperature and would delay the high-temperature transition so much that significant damage would occur before the measured temperature triggered an alarm or shutdown (other interlocks such as high thrust would activate long before high gas temperature).

8. Magnetic pickups would require the shaft to be magnetic and would be wiped out during large amplitude radial vibration or normal rotor axial movement. The linear span requirement for thrust measurement is far beyond the span capability of the magnetic pickup.

Section 8

1. The measurement noise would decrease the allowable controller gain, which would increase the peak error per Equation 8-1.

2. The quick-opening installed flow characteristic would cause the operating point to slowly approach the set point, keeping the surge valves closed, and then rapidly approach the surge curve. The surge

valves may not open in time to prevent surge. The linear approach to the surge set point and the surge curve is better for surge control.

Section 9

1. A throughput control loop would normally be much slower than a properly designed surge control loop. The throughput loop's controller gain would be smaller and period would be larger so that decoupling is unnecessary.
2. A mass balance override or 'anticipator' control system opens the surge valve for disturbances far away from the computer surge set point. Whenever the surge valve is throttled, interaction occurs.

Section 10

1. The delay of the correction by a downstream surge control valve increases as the separator volume increases because the pressure must first drop in the separator before the flow upstream into the separator will increase. The upstream compressor is more likely to go into surge.
2. The leader should be selected based on proximity to the surge curve in order to minimize unnecessary venting or recycle and to maximize the distance of all compressor operating points from the surge curve.

Section 11

1. The real time analyzer computes the Fast Fourier Transform for the computer so that an array processor in the computer is not required. Array processors are not generally available in process control computers.
2. Since the flow drops to its minimum in less than 0.05 second at the start of surge, a transmitter time constant or computer sample time slower than this will use the measurements at the bottom instead of at the start of the surge cycle as the coordinates of the surge point.

Section 12

1. The control valve is typically the slowest instrument in the loop and thus affects the controller gain and integral time settings the most. Since surge control valves typically have large pneumatic actuators, it is difficult to make them as fast as the electronic instruments in the loop.

2. The surge controller becomes a minimum flow controller (see Section 6-1) if the measurement is suction flow. It becomes a maximum pressure controller if the measurement is pressure rise (see Section 6-2). Both types of controllers provide surge protection interior to the remote set point surge controller.

ACSL Dynamic Simulation Program for Surge

PROGRAM DYNAMIC MODEL OF SURGE MAP (ACSL FUNCTIONAL BLOCK)
'MODEL IS FROM P 391–395 OF SEPT 1981 JOURNAL OF FLUIDS ENGINEERING'
'Y(1) = COMPRESSOR INLET VOLUMETRIC FLOW (ACFS)'
'Y(2) = THROTTLE VALVE INLET VOLUMETRIC FLOW (ACFS)'
'Y(3) = SURGE VALVE INLET VOLUMETRIC FLOW (ACFS)'
'Y(4) = DYNAMIC PRESSURE RISE FROM SUCTION TO PLENUM (PSI)'
'Y(5) = DYNAMIC PRESSURE RISE IN THE COMPRESSOR (PSI)'
'Y(6) = STEADY STATE PRESSURE RISE IN THE COMPRESSOR (PSI)'
'Z(1) = DIMENSIONLESS COMPRESSOR INLET FLOW'
'Z(2) = DIMENSIONLESS THROTTLE VALVE INLET FLOW'
'Z(3) = DIMENSIONLESS SURGE VALVE INLET FLOW'
'Z(4) = DIMENSIONLESS DYNAMIC PRESSURE RISE FROM SUCTION TO PLENUM'
'Z(5) = DIMENSIONLESS DYNAMIC PRESSURE RISE IN THE COMPRESSOR'
'Z(6) = DIMENSIONLESS STEADY STATE PRESSURE RISE IN THE COMPRESSOR'
'Z(7) = DIMENSIONLESS STABILITY PARAMETER B'
'Z(8) = MINIMUM VALUE OF PARAMETER B FOR STABILITY'
'Z(9) = MAXIMUM SURGE FREQUENCY (HELMHOLTZ FREQUENCY) (CYCLES/SEC)'
'P(1) = SPEED OF SOUND AT SUCTION CONDITIONS (FT/SEC)'
'P(2) = COMPRESSOR FLOW AREA (SQFT)'
'P(3) = THROTTLE VALVE FLOW AREA (SQFT)'
'P(4) = SURGE VALVE FLOW AREA (SQFT)'
'P(5) = LENGTH OF THE COMPRESSOR (FT)'
'P(6) = LENGTH OF THE PLENUM (FT)'
'P(7) = LENGTH OF THE THROTTLE VALVE PIPING (FT)'

'P(8) = LENGTH OF THE SURGE VALVE PIPING (FT)'
'P(9) = TIME LAG IN REVOLUTIONS'
'P(10) = IMPELLER TIP RADIUS (FT)'
'P(11) = PLENUM VOLUME (CUFT)'
'P(12) = DENSITY OF THE GAS AT SUCTION CONDITIONS (LBF/CUFT)'
'P(13) = CHARACTERISTIC CURVE NEGATIVE FLOW BRANCH POLYNOMIAL A0 COEFF'
'P(14) = CHARACTERISTIC CURVE NEGATIVE FLOW BRANCH POLYNOMIAL A1 COEFF'
'P(15) = CHARACTERISTIC CURVE NEGATIVE FLOW BRANCH POLYNOMIAL A2 COEFF'
'P(16) = CHARACTERISTIC CURVE UNMEASURED BRANCH POLYNOMIAL A0 COEFF'
'P(17) = CHARACTERISTIC CURVE UNMEASURED BRANCH POLYNOMIAL A1 COEFF'
'P(18) = CHARACTERISTIC CURVE UNMEASURED BRANCH POLYNOMIAL A2 COEFF'
'P(19) = CHARACTERISTIC CURVE UNMEASURED BRANCH POLYNOMIAL A3 COEFF'
'P(20) = CHARACTERISTIC CURVE STABLE BRANCH POLYNOMIAL A0 COEFF'
'P(21) = CHARACTERISTIC CURVE STABLE BRANCH POLYNOMIAL A1 COEFF'
'P(22) = CHARACTERISTIC CURVE STABLE BRANCH POLYNOMIAL A2 COEFF'
'P(23) = CHARACTERISTIC CURVE STABLE BRANCH POLYNOMIAL A3 COEFF'
'P(24) = CHARACTERISTIC CURVE STABLE BRANCH START POINT'
'P(13) to P(24): DIMENSIONLESS COMPRESSOR MAP COORDINATES Z(1) & Z(6)'
'Z(1) = ABSCISSA: DIMENSIONLESS COMPRESSOR FLOW CALCULATED BY DIVIDING'
'THE VELOCITY AT THE SUCTION BY THE IMPELLER TIP SPEED EACH IN FT/SEC'
'Z(6) = ORDINATE: DIMENSIONLESS COMPRESSOR PRESSURE RISE CALCULATED BY'
'DIVIDING THE PRESSURE RISE IN LB/SQFT BY ONE HALF THE PRODUCT OF THE'
'DENSITY IN LB/SQFT AND THE SQUARE OF THE IMPELLER TIP SPEED IN FT/SEC'
'X(1) = COMPRESSOR SPEED (RPM)'
'X(2) = THROTTLE VALVE THROAT AREA (SQFT)'
'X(3) = SURGE VALVE THROAT AREA (SQFT)'
INITIAL
ARRAY P(24), Y(6), Z(9), ZIC(5), X(3)
MACRO SURGE (Y, Z, ZIC, P, X)
MACRO REDEFINE B, F, G, H, I, J, U, ZD, ZI
ARRAY ZD(5), ZI(5)
U=2.*3.14*P(10)*X(1)/60.
B=(U/(2.*P(1)))*SQRT(P(11)/(P(2)*P(5)))
F=(Z(2)*P(2)/X(2))**2
G=(P(7)/P(3))/(P(5)*P(2))
H=(Z(3)*P(2)/X(3))**2
I=(P(8)/P(4))/(P(5)*P(2))
J=(3.14*P(9)*P(10))/(P(5)*B)
PROCEDURAL (ZD=Z,B,F,G,H,I,J)
ZD(1)=(Z(5)−Z(4))*B
ZD(2)=(Z(4)−F)*B/G
ZD(3)=(Z(4)−H)*B/I
ZD(4)=(Z(1)−Z(2)−Z(3))/B
ZD(5)=(Z(6)−Z(5))/J
END

```
ZI=INTVC(ZD,ZIC)
PROCEDURAL (Z,Y=P,X,B,U,ZI)
Z(1)=ZI(1)
Z(2)=ZI(2)
Z(3)=ZI(3)
Z(4)=ZI(4)
Z(5)=ZI(5)
Z(6)=P(20)+P(21)*Z(1)+P(22)*Z(1)**2+P(23)*Z(1)**3
IF (Z(1).LT.P(24)) Z(6)=P(16)+P(17)*Z(1)+P(18)*Z(1)**2+P(19)*Z(1)**3
IF (Z(1).LT.0.000) Z(6)=P(13)+P(14)*Z(1)+P(15)*Z(1)**2
Z(7)=B
DFDM=2.*P(24)*(P(2)/X(2))**2
DCDM=P(17)+2.*P(18)*P(24)+3.*P(19)*P(24)**2
Z(8)=SQRT(I./(DFDM*DCDM))
IF (Z(8).GT.I.0) Z(8)=I.0
Z(9)=P(1)*SQRT(P(2)/OP(5)*P(11)))/6.28
Y(1)=Z(1)*U*P(2)*60.
Y(2)=Z(2)*U*P(3)*60.
Y(3)=Z(3)*U*P(4)*60.
Y(4)=(Z(4)*0.5*P(12)*U**2)/(144.*32.2)
Y(5)=(Z(5)*0.5*P(12)*U**2)/(144.*32.2)
Y(6)=(Z(6)*0.5*P(12)*U**2)/(144.*32.2)
END
MACRO END
CONSTANT ZIC=0.3,0.3,0.0,0.9,0.9
CONSTANT P=1140.,0.036,0.036,0.036,4.0,8.2,2.6,2.6,0.5,0.15,0.88,...
    0.077,0.85,0.0,21.9,0.85,0.0,58.0,-254.,1.4,-3.0,25.6,-69.0,0.152
CONSTANT X=54000.,0.036,0.00036
CONSTANT TSD=0.,TFD=0.99
CONSTANT TSTOP=24.
ALGORITHM IALG=1
CINTERVAL CINT=0.02
END
DYNAMIC
DERIVATIVE
Y,Z=SURGE(ZIC,P,X)
TIME=T/(6.28*Z(9))
PROCEDURAL(X(2)=TSD,TFD)
X(2)=0.036*(1.0-RAMP(TSD)+RAMP(TSD+TFD))
END
ACFM=Y(1)
PSI=Y(4)
PATH=Y(4)
CURVE=Y(6)
END
```

```
TERMT (T.G.E.TSTOP)
END
TERMINAL
END
END
SET PRN=9
PREPAR TIME,Y,Z,ACFM,PSI,PATH,CURVE
OUTPUT TIME,ACFM,PSI,'NCIOUT'=10
PLTDIN
SET XINCPL=8.,YINCPL=6.,NPCCPL=50
SET TTLCPL=.T.
PROCED RUN
START
DISPLY TIME, ACFM, PSI
DISPLY Z(7), Z(8), Z(9)
SET TITLE ='PLOT 1 IS THE FLOW (ACFM VS TIME)    PLOT 2 IS THE...PRESSURE (PSI VS
    TIME)
PLOT 'XASIS'=ACFM,'XLO'=−300.,'XHI'=600.,PATH,'CHAR'='1',...END
RUN
SET TITLE ='PLOT 1 IS THE PATH (PSI VS ACFM    PLOT 2 IS THE...CURVE (PSI VS ACFM)
PLOT 'XAXIS'=ACFM,'XLO'= ⅞ 300.,'XHI'=600.,PATH,'CHAR'='1',...CURVE,'CHAR'='2'
SET X(1)=9000.
RUN
SET TITLE ='PLOT 1 IS THE PATH (PSI VS ACFM)  PLOT 2 IS THE...CURVE (PSI VS ACFM)
PLOT 'XAXIS'=ACFM,'XLO'=−80.,'XHI'=100.,PATH,'CHAR'='1',CURVE,'CHAR'='2'
SET X(1)=110.
RUN
SET TITLE ='PLOT 1 IS THE PATH (PSI VS ACFM)  PLOT 2 IS THE...CURVE (PSI VS ACFM)
PLOT 'XAXIS'=ACFM,'XLO'=0.,'XHI'=1.2,PATH,'CHAR'='1',CURVE,'CHAR'='2'
SET CMD=5
```

Application Example

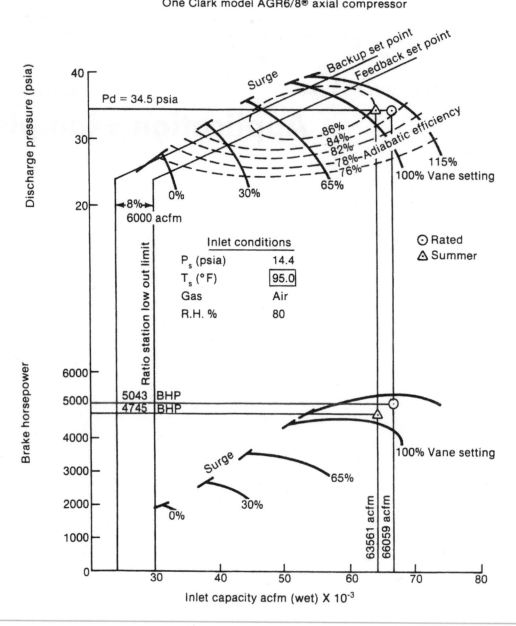

Figure C-1   Example of Compressor Map (Summer Operation)

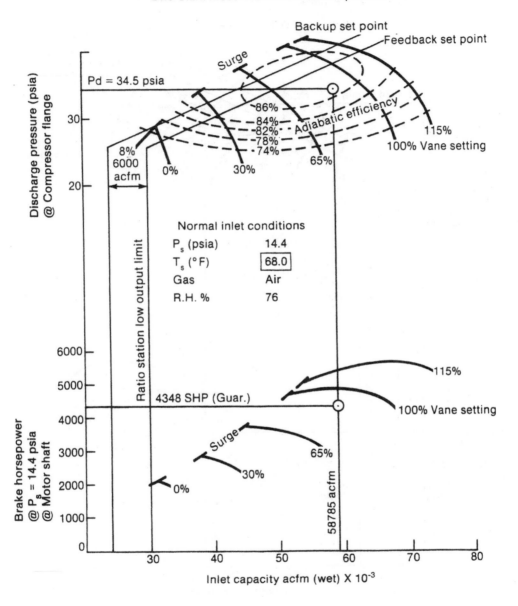

Figure C-2   Example of Compressor Map (Spring Operation)

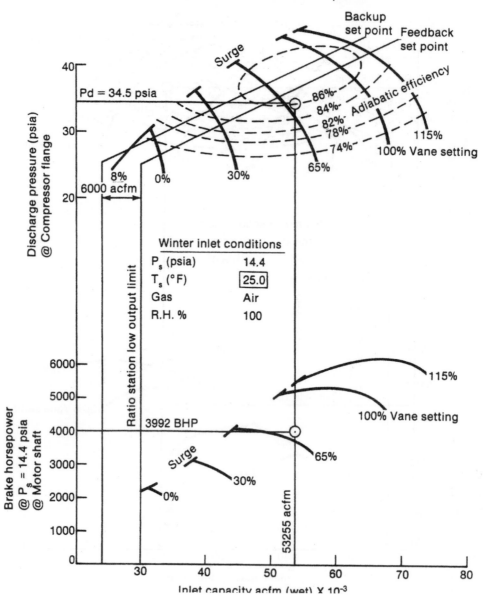

Figure C-3   Example of Compressor Map (Winter Operation)

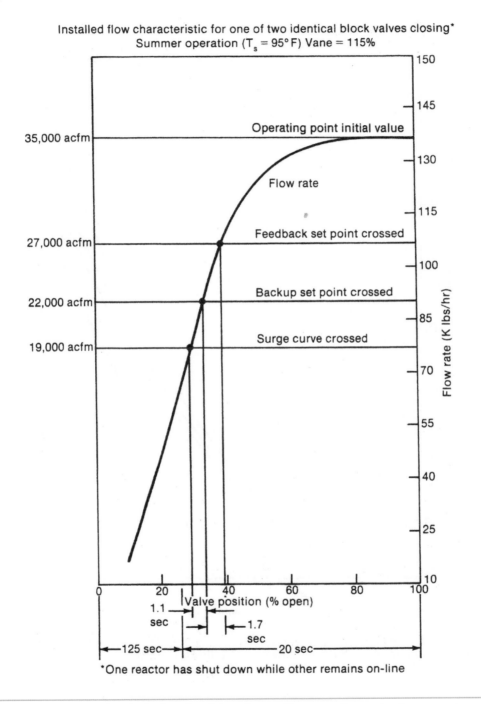

Installed flow characteristic for one of two identical block valves closing*
Summer operation (T$_s$ = 95° F) Vane = 115%

Operating point initial value

Flow rate

Feedback set point crossed

Backup set point crossed

Surge curve crossed

35,000 acfm

27,000 acfm

22,000 acfm

19,000 acfm

Flow rate (K lbs/hr)

Valve position (% open)

1.1 sec

1.7 sec

125 sec

20 sec

*One reactor has shut down while other remains on-line

Figure C-4   Example of Block Valve Disturbance (Summer Operation)

Clark model AGRG/8® axial compressor calibration
curve for inlet nozzle flow measurement

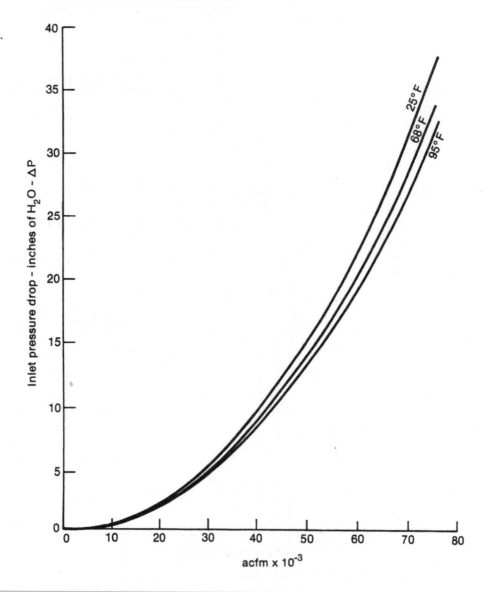

Figure C-5   Example of Suction Flowmeter Calibration Shift

References

1. American Petroleum Institute Standard 670.
2. AVCO 7600 SAFE System Bulletin, "Signature Analysis," 1980.
3. Baker, D.F., "Surge Control for Multistage Centrifugal Compressors," *Chemical Engineering*, p. 117–122, May 31, 1982.
4. Bently Nevada, "Probe and Proximitor Installation Manual—Section 1 Theory of Operation," 1973.
5. Bently Nevada, "Application Manual for Digital Tachometers," 1973.
6. Bently Nevada, "Orbits," 1973.
7. Catheron, A.R., "A Method of Reducing Turbulence Noise Effects in Flow Control," ISA First Symposium on Flow, Pittsburgh, 1971.
8. Boyce, M.P., "How to Achieve Online Availability of Centrifugal Compressors," *Chemical Engineering*, p. 115–127, June 5, 1978.
9. Dieck, R.H., et. al., "Thermocouple Measurement Uncertainty in Compressor Efficiency Measurement: The Effects of Two Uncertainty Models," *Temperature—Its Measurement and Control in Science and Industry*, Volume 5, American Institute of Physics, p. 1009–1018, 1982.
10. Dimoplan, W., "What Process Engineers Need to Know about Compressors," *Hydrocarbon Processing Magazine Compressor Handbook*, p. 1–8, 1979.
11. Dwyer, J.J., "Compressor Problems: Causes and Cures," *Hydrocarbon Processing Magazine Compressor Handbook*, p. 9–12, 1979.
12. Gaston, J.R., "Centrifugal Compressor Operation and Control—Part II: Compressor Operation," ISA 31st Annual Conference, Houston, 1976.
13. Gaston, J.R., "Improved Flow/Delta-P Anti-Surge Control System," Pacific Energy Association, Compressor-Engine Workshop, p. 1–13, March 1981.
14. General Electric MDT-80 Electro-Hydraulic Control System Bulletin.
15. Greitzer, E.M., "The Stability of Pumping Systems—The 1980 Freeman Scholar Lecture," *Journal of Fluids Engineering*, p. 193–242, June 1981.
16. Hansen, K.E. et al., "Experimental and Theoretical Study of Surge in a Small Centrifugal Compressor," *Journal of Fluids Engineering*, p. 391–395, Sept. 1981.

17. Harker, R.G., "A Dedicated Mini-computer-Based Monitoring System for Turbomachinery Protection and Analysis," ASME 76-Pet-13, 1976.

18. Jackson, C., "How to Prevent Turbomachinery Thrust Failures," *Hydrocarbon Processing Magazine Compressor Handbook*, p. 152–157, 1979.

19. Kerlin, T.W. et al., "Response of Installed Temperature Sensors," *Temperature—Its Measurement and Control in Science and Industry*, Volume 5, American Institute of Physics, p. 1009–1018, 1982.

20. Mamzic, C.L., "Improving the Dynamics of Pneumatic Positioners," *ISA Journal*, p. 38–43, Aug. 1958.

21. McMillan, G.K., *Tuning and Control Loop Performance*, Momentum Press, 2010.

22. Pollack, R., "Selecting Fans and Blowers," *Chemical Engineering*, p. 86–100, Jan. 22, 1973.

23. Rammler, R., "Centrifugal Compressor Performance Control," Powell Process Systems, Inc., Technical Paper.

24. Staroselsky, N. and Ruthstein, A., "Some Considerations on Improving the Control Strategy for Dynamic Compressors," *ISA Transactions*, Vol. 16 No. 2, p. 3–19, 1977.

25. Staroselsky, N., "Improved Surge Control for Centrifugal Compressors," *Chemical Engineering*, p. 174–184, May 21, 1979.

26. Staroselsky, N., "Better Efficiency and Reliability for Dynamic Compressors Acting in Series or Parallel," ASME Petroleum Division paper at Energy Technology Conference, Feb. 1980.

27. Stryker, J.E., "Basic Specification Considerations," *Hydrocarbon Processing Magazine Compressor Handbook*, p. 31–36, 1979.

28. Uram, R., "Hardware and Software Team Up to Control Steam Turbine Speed," *Control Engineering*, Feb. 1976.

29. White, M.H., "Surge Control for Centrifugal Compressors," *Chemical Engineering*, Dec. 25, 1972.

30. McMillan, G.K., *Essentials of Modern Measurements and Final Elements in the Process Industry—A Guide to Design, Configuration, Installation, and Maintenance*, ISA, 2010.